T0213766

Harnessing the Power of Analytics

Leila Halawi • Amal Clarke • Kelly George

Harnessing the Power
of Analytics

Leila Halawi
College of Aeronautics
Embry–Riddle Aeronautical University
Clearwater Beach, FL, USA

Amal Clarke
College of Arts and Science
Embry–Riddle Aeronautical University
Frimley, Surrey, UK

Kelly George
College of Aeronautics
Embry–Riddle Aeronautical University
South Riding, VA, USA

ISBN 978-3-030-89714-7   ISBN 978-3-030-89712-3   (eBook)
https://doi.org/10.1007/978-3-030-89712-3

This Springer imprint is published by the registered company Springer Nature Switzerland AG
The registered company address is: Gewerbestrasse 11, 6330 Cham, Switzerland

Contents

1   Introduction to Analytics and Data Science   1
   1.1   The Data Analytics and Data Science Revolution   1
   1.2   The Difference Between Data Analytics and Data Science   2
   1.3   Data Analyst Versus Data Scientist   3
   1.4   Example: Data Science and Analytics in Aviation   4
   1.5   Analytics Methods   5
   1.6   Classification of Different Applications and Vendors   7
   1.7   Why the Many Different Methods?   8
   1.8   What You Need to Know About SAS Viya   8
   1.9   Road Plan for the Book   9
   References   10

2   Data Types Structure and Data Preparation Process   13
   2.1   Data Types and Measurements   13
      2.1.1   Taxonomy Data   13
      2.1.2   Big Data   14
   2.2   The Need for Data and Data Sources   15
   2.3   Data Sources   15
   2.4   Data Partitioning and Honest Assessment   16
   2.5   The Necessity of Data Preparation and Curation   17
   2.6   Data Preparation Process   17
   2.7   Exploring SAS VIYA Plaftorm and Import Data   21
   2.8   Understanding the SAS VIYA Interface   21
   2.9   Loading Data into SAS VIYA   23
   Reference   27

3   Data Exploration and Data Visualization   29
   3.1   Aviation Case   29
   3.2   Investigate Phases of Data and Data Exploration   31
   3.3   Putting Descriptive Measures Together   32
   3.4   Descriptive Statistics   33
      3.4.1   Measures of Central Tendency   34

   3.4.2 Measures of Variation 34

   3.4.3 Distribution Shapes 36

  3.5 Data Visualization 38

   3.5.1 Data Visualization Considerations 38

   3.5.2 Types of Data Visualization 40

  3.6 Application Using SAS VIYA 43

  3.7 Answers to Aviation Case Questions 45

  References 50

4 Evaluating Predictive Performance 51

  4.1 The Importance of Evaluating Predictive Performance 51

  4.2 Algorithms 52

  4.3 Evaluating Predictive Performance 54

   4.3.1 Model Assessment 54

   4.3.2 Metrics 54

   4.3.3 Selecting Model Fit Statistics by Prediction Type 57

  References 59

5 Decision Trees and Ensemble 61

  5.1 Overview of Decision Trees 61

   5.1.1 Classification Trees 62

   5.1.2 Regression Trees 62

  5.2 Terms Used with Decision Trees 63

  5.3 The Math Behind Decision Trees 63

   5.3.1 Expected Value 63

   5.3.2 Measure of Level of Impurity 64

   5.3.3 Attribute Selection Measures 64

  5.4 Avoiding Overfitting 70

   5.4.1 Pruning 71

   5.4.2 Concrete Compressive Strength Example 1: Regression Tree

     Model SAS Visual Analytics 72

  5.5 Ensemble Methods Explained 76

  5.6 Ensemble Methods 77

   5.6.1 Bagging or Bootstrap Aggregation 77

   5.6.2 Boosting 79

  References 81

6 Regression Models 83

  6.1 Functions and Mathematical Implementation 83

   6.1.1 Functions 83

   6.1.2 Coordinate Plane 84

   6.1.3 Derivative of Functions 85

   6.1.4 Matrices 86

   6.1.5 Definition of Logarithmic Function 88

6.2 Linear Regression 88
    6.2.1 Applications of Linear Regression. 91
6.3 Multiple Linear Regression 91
    6.3.1 Estimation of the Model Parameters 93
    6.3.2 Concrete Compressive Strength Example 1: SAS Visual
         Analytics 94
    6.3.3 Concrete Compressive Strength Example 1: Model Builder 99
6.4 Logistic Regression 101
    6.4.1 Estimating the Coefficients 104
    6.4.2 Types of Logistic Regression........................ 105
    6.4.3 Demonstrations of Logistic Regression: Aviation Example 106
Reference 108

7 Neural Networks 109
7.1 What Are Neural Networks? 109
    7.1.1 How Do Neural Networks Learn? 110
7.2 The Architecture of Neural Networks 111
    7.2.1 Terminology...................................... 111
7.3 The Mathematics Behind Neural Network 113
    7.3.1 The Common Activation Functions................... 113
    7.3.2 Limit of Functions 114
    7.3.3 Chain Rule....................................... 114
    7.3.4 Working of Neural Network 115
    7.3.5 Backward Propagation............................. 117
7.4 Vanishing and Exploding Gradient 119
7.5 Demonstration for Neural Networks 120
    7.5.1 Concrete Compressive Strength Example 1: SAS Visual
         Analytics 120
    7.5.2 Demonstrations of Neural Network: Aviation Example ... 122
References.. 127

8 Model Deployment 129
8.1 Model Deployment 129
    8.1.1 Model Assessment 130
    8.1.2 Model Comparison. 130
    8.1.3 Monitoring Model Performance and Updating 132
8.2 Advantages and Disadvantages of Decision Trees, Regression, Neural
    Networks, Forest, and Ensemble Models. 132
8.3 Application Example of Model Deployment: Concrete Compressive
    Strengths 133
8.4 Conclusion.. 135
References.. 136

Appendix A: Information for Instructors 137

Appendix B: Sources of Public Data 145

Appendix C: Data Dictionary for the Aviation......................... 147

Appendix D: Data Dictionary for the Concrete 149

Chapter 1
Introduction to Analytics and Data Science

Learning Objectives

- Grasp the difference between and need for analytics and data science
- Recognize the difference between the data analyst and data scientist
- Identify and describe the different types of analytics
- Classification of different applications and vendors
- Identify the many different methods of prediction
- Introducing SAS Viya
- Road plan for the book

1.1 The Data Analytics and Data Science Revolution

The future pauses for no one. Change nowadays is more complicated, quick, and tough to predict than ever. Data have and will continue to transform businesses such as FedEx, Google, Intel, Apple, Tesla, Uber, and Amazon. Innovation driven by data, powered by venture capital, is restructuring the world. The Internet provides instantaneous access to just about all types of information, and we should anticipate similar urgency for all kinds of solutions within the workplace. Data operationalization and manipulation enhance business functioning to best define competitive advantage well into the future. This voracious desire for data is a cultural change and paradigm shift witnessed on a global stage (Kiron and Shockley 2011).

Data analytics and data science are now being promoted in online media, journals, and even at conferences, thus surpassing the limited time to a publication that printed books present. Data science and analytics have the power to predict political races in real time, expose buying habits, and predict, with succinct results, many of our pressing research questions today. According to a study by Gartner (2018), a worldwide survey ($n = 196$) established that 91% of companies

© The Author(s), under exclusive license to Springer Nature
Switzerland AG 2022
L. Halawi et al., *Harnessing the Power of Analytics*,
https://doi.org/10.1007/978-3-030-89712-3_1

have not yet achieved a "transformational" maturity level in data science and analytics. Therefore, data science and analytics are now the number 1 investment priority of CIOs in recent years (Meulen and McCall 2018). Historically, novel technological advances initially emerged in technical and academic periodicals. The knowledge and synthesis presently seeped into other publications, many in book format.

Data science and the related fields of business intelligence and analytics are becoming progressively central to academic and business communities, as seen in recent literature (Chen et al. 2012). However, there is still a great deal of confusion surrounding the meaning of data science and analytics among practitioners and academia.

To dispel this confusion, especially for those seeking a career in these fields, one needs a clear definition of data analytics and data science and then a clear occupational description of the job of a data analyst as opposed to the data scientist, including the duties, tasks, knowledge, skills, and traits.

1.2   The Difference Between Data Analytics and Data Science

While data analytics and data science are often used as synonyms, there are differences between the two disciplines, most specifically in terms of knowledge and skills.

The phrase data analytics is used in place of business intelligence (BI). Many practitioners and consultants describe analytics differently. For the Institute for Operation Research and Management Science (INFORMS), analytics denotes the mixture of computer technology, management science techniques, and statistics to resolve original problems. Of course, other enterprises propose their interpretation and motivation for analytics. For SAS Institute Inc., analytics follows the data everywhere, and analytics is more than algorithms; the value generated is the focus point (Schabenberger 2020). Throughout this book, "data analytics" will generally refer to the analysis of data sets to derive meaningful trends and develop visual displays of existing data to help businesses answer questions or problems.

Data science is not a specific discipline in itself but yet spans across all disciplines and is a broader term than Data Analytics. Data science is a ubiquitous term designating an interdisciplinary field regarding processes and systems to gain knowledge and insights from data (Bichler et al. 2017). Data science generally refers to generating information from large unstructured and structured data sets. Figure 1.1 from Northeastern University highlights the difference between the disciplines, the methods, and the overlap (Burnham 2019).

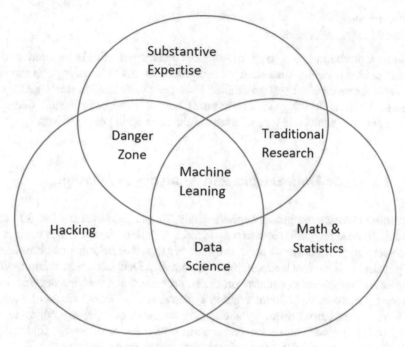

Fig. 1.1 Data analytics and data science skills

1.3   Data Analyst Versus Data Scientist

Data analytics and data science fields offer a range of duties, tasks, knowledge, skills, and traits. There is a clear overlap between the data scientist and the data analyst. However, the breadth and depth of the skills necessary differ, making each role complementary but different in focus, viewpoint, and expertise.

A data analyst is a professional who used to do "business intelligence" (now called analytics) in data compilation, cleaning, reporting, and visualization. Some of these professionals may have a more in-depth knowledge of programming to code for data cleaning and analysis.

A "talented" data scientist would generally be supposed to have rigorous business skills to assess the significance of generated insights and address significant business problems (Provost and Fawcett 2013).

According to NAP (2014), extracting meaning from data requires considerable skills that include:

- Statistics
- Machine learning
- Optimization
- Software engineering

- Product sense
- Careful experimentation

While a profound appreciation of numbers and mathematics is imperative to be successful in this field, a data scientist must also possess outstanding communication skills, be an eminent system thinker, have keen visualization skills, and above all possess critical thinking skills concerning how data can be utilized in decision-making and the impact this type of analysis can have upon people's lives.

1.4 Example: Data Science and Analytics in Aviation

The airline industry has grown approximately 5% per year over the last 30 years. The data science revolution is also transforming aviation. From airlines to air navigation service providers or airports, the capacity to gather information across separate physical data sensors is expanding exponentially. Data science in aviation offers countless opportunities to enhance products, processes, and imagine new means to develop safer and more efficient aviation systems. A significant amount of unstructured, varied data from distinct stakeholders and various types are collected and stored within the aviation sector, comprising safety data and reports, flight plans, navigation data, airport data, and radar tracks, among other data types.

Airlines and airports have limited capability to process this haul of data and employ advanced analytics and artificial intelligence to inform operations and maintenance and rarely in real time (Maire and Spafford 2017). Airlines can use data to fine-tune their fuel loads, beverage inventory, and personnel requirements to save money and increase efficiency. To bridge the gap between supply and demand, training the next generation, and retooling the current aviation workforce is necessary for the long run. Yet, it will not suffice long term; it is not sufficient.

The Bureau of Aircraft Accident Archives (B3A) reports that, on average, 230 fatal aviation accidents occurred, and 1709 passages perished annually in the last half-century. The best estimate we found from the literature states that "somewhere between 60 and 80% of aviation accidents are due, at least in part, to human error" (Shappell 2006). Such a preliminary estimate can be and should be improved at the current age of data science and the state of artificial intelligence (AI) technology. However, pinning down the exact measures for reducing such tragic accident rates and annoying flight delays need thousands of data points to analyze the ground truth of enormous aviation data.

The subject matter expertise in aviation is extensive enough for a non-aviation background person to understand and interpret most aviation data sets. The aviation industry mostly hires students with an aviation background paired with a data management perspective.

Data science and analytics in aviation offer countless opportunities to enhance products, processes, and imagine new means to develop safer and more efficient aviation systems. Within the aviation sector, a significant amount of unstructured, varied data from distinct stakeholders and various types is collected and stored, comprising safety data and reports, flight plans, navigation data, airport data, and radar tracks, among other data types. A recent *Aviation Week* article highlights the characteristics of the new Pratt & Whitney Geared TurboFan (GTF) engines that employ 5000 sensors and are capable of generating up to 10 GB of data per second. By comparison, current engines only have 250 sensors at most. This results in a twin-engine aircraft equipped with the new Pratt & Whitney GTF, with an average of 12 h flight, generating up to 844 TB of data. With thousands of engines to be built, zettabyte (ZB) of data will be available.

Consequently, the infrastructure needed to handle such data needed significantly upgraded. That infrastructure will be required to be put in place to benefit from the wealth of information the engine data could provide. From this point of view, engine health monitoring is getting an entirely new perspective. The INNAXIS research institute deems the application of data science principles to the aviation sector as an open gate to substantial improvements in numerous main aspects of aviation, such as safety enhancement, flight efficiency, environmental impact mitigation, or delay reduction. According to Maire and Spafford (2017), the flight-related data amount is increasing significantly, and this increase enables making a more profound analysis; however, airlines and airports have limited capability to process this haul of data and employ advanced analytics and AI to inform operations and maintenance and rarely in real time.

SAS has been successfully used by several airlines for various use cases. See the SAS website for more information (www.sas.com/fi_fi/customers/scandinavian-airlines.html).

1.5 Analytics Methods

INFORMs, The Institute for Operations Research and the Management Sciences, proposed three levels of analytics. These levels are identified as descriptive, predictive, and prescriptive (Informs 2014).

Figure 1.2 presents a graphical overview of the analytics process with the three different analytics levels.

While descriptive analytics focuses primarily on what has already happened in the past, and predictive analytics tries to find correlations to make forward-looking projections, prescriptive analytics looks to determine what to do or give you an answer as to how to proceed on the information provided to the data model.

1. **Descriptive analytics—What happened and what is happening?** Descriptive analytics is the entry-level in the analytics world. It frequently entails working with queries, looking at descriptive statistics, data visualization including dashboards, and generating reports as most of the analytics actions at this level involve creating a report to summarize business activities and answer the question, "What happened or what is happening?" For example, a query to a database

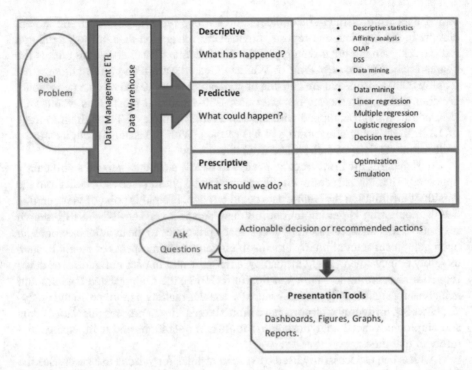

Fig. 1.2 Analytics process and levels

for a shipment facility for Amazon during April will provide descriptive information about all the shipments, the number of purchases, the dates for each
delivery, and the type and quantities shipped per order, among other things. A
report can later be generated with summary statistics such as mean, standard
deviation, and some data visualizations to detect any patterns or relationships.

2. **Predictive analytics—What will happen? Why will it happen?** Organizations
that matured in descriptive analytics often move into this level to look beyond
what happened and try to predict or answer what will happen in the future. This
form of analysis is usually based on statistical techniques and, data mining
techniques. For example, in the financial sector, predictive models predict future
financial performance and assess investments' risks. Another example used
widely across all retail companies and services providers in developing predictive models to support pricing decisions they take, understand consumer behavior, increase loyalty, and customer satisfaction, among other uses.

3. **Prescriptive analytics**—What is the best outcome that will happen? Prescriptive
analytics is the ultimate level in the analytics world or hierarchy. It is where the
most excellent choice among many created and classified is determined using
sophisticated mathematical models. This type of analysis answers the question:
what should I do? It uses optimization, simulation, and heuristics-based decision-making modeling techniques. This group of methods has traditionally been
studied under operation research (OR) or management sciences. For example, in

Table 1.1 The three forms of analytics

Descriptive analytics	Also known as exploratory descriptive analytics. Here you detect data inconsistency and formatting issues and execute statistical tests to establish which variables appear most closely related, complete, useful, and incomplete. Exploratory descriptive analytics is commonly aimed to provide insights into the variations, relationships, and patterns in data that can inform later analyses Descriptive analytics may also be focused on answering specific business questions. What were the sales last month? What was the shortage in a given product? It is important to note that the only thing that descriptive analytics requires is data (and a tool to analyze it)
Predictive analytics	The purpose of predictive analysis is to determine what is likely to happen in the future based on data mining techniques Running a regression to explain the impact of $X1$ of Y is descriptive. Building a predictive model with training and validation data sets to predict and score future Y data based on $X1$ is predictive In these models, we identify the who, what, when, where, and so on, which is vital
Prescriptive analytics	Its goal is to anticipate what is going well and the likely forecast to realize the best performance. For example, optimization tells you what to do (how many of each product to produce based on the expected demand of each product, which products to launch to market based on consumer survey responses, etc.) Identifying the *why* improves understanding the who, what, when, where, and so on, and the different ways they are connected

the sports industry, companies use prescriptive analytics to dynamically adjust their ticket prices throughout the season, reflecting each game relative attractiveness and potential demand.

Predictive and prescriptive analytics are collectively called advanced analytics. Prescriptive analytics may be considered the last step of business analytics. However, it is essential to note the difference between these two forms related to the outcome of the analysis. Also, predictive analytics offers you data in hard ways to help you make informed decisions.

For example, healthcare uses descriptive, predictive, and prescriptive analytics to improve the facility, staff, and patients' scheduling, provide more effective treatments, predict patient flow and diagnosis, and even treatments.

Table 1.1 summarizes the three forms of analytics.

1.6 Classification of Different Applications and Vendors

Gartner, Inc., formerly known as Gartner, is a worldwide research and advisory firm offering information, advice, and tools for leaders in IT, customer service and support, and supply chain functions, among others. The Magic Quadrant 2020 assesses data analytics and BI platform vendors and data science and machine learning (DSML) platforms.

Richardson et al. (2021) identified many different vendors within the data analytics and BI platforms and compared them across many different functionalities and capabilities. These included security, cloud, data source connectivity, manageability, data preparation, data visualization, reporting, advanced analytics, and model complexity, to name a few. Microsoft, Tableau, SAS, SAP, Qlik, Microstrategy, Oracle, IBM, and others were identified, and advantages and disadvantages for each platform were listed.

The data science and machine learning (DSML) platforms, Gartner (2020a, b) reviewed and classified different vendors based on various features including data exploration, data preparation, model testing, deployment, maintenance, and collaboration. They also presented the distinct advantages and disadvantages for each vendor. This information is an integral part of the selection process for any organization. Vendors such as Alteryx, SAS, Databricks, MathWorks, Tibco Software, and Dataiku were classified as leaders in the field.

1.7   Why the Many Different Methods?

There are many prediction methods and one cannot but ask why they coexist and whether some are better than others. One should know that each method has advantages and disadvantages.

The data set's size may impact the usefulness of a particular method, the analysis goal, how messy the data are, the existing patterns, if any, in the data set, and whether the data meets the method's underlying assumptions. Figure 1.3 highlights the different techniques across the data analytics and data science field and the need for application tools.

1.8   What You Need to Know About SAS Viya

SAS has evolved a lot in the last 30 years and is still among the elites in the artificial intelligence (AI) and advanced analytics world. SAS® Viya® is an open analytics platform that may process any data type, volume, or speed. This cloud-enabled, in-memory analytics engine is adaptable, scalable, and fault-tolerant. It includes a standardized code base that adopts programming in SAS and other languages, such as Python, R, Java, and Lua. Furthermore, it may deploy flawlessly to any infrastructure or product ecosystem with support for cloud, on-site, or hybrid environments. The high-performance processing power of SAS Viya is offered by SAS Cloud Analytics Services (CAS), an in-memory engine that can radically hasten data management and analytics with SAS. SAS Viya is planned to synchronize with SAS 9.4 solutions and the SAS 9 environment. Some procedures are available in both SAS 9 and SAS Viya. While some existing SAS code can still run in SAS Viya,

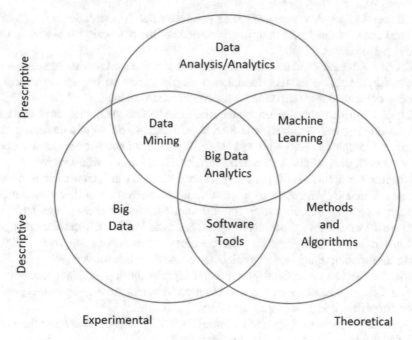

Fig. 1.3 The fields of data analytics and data science

SAS Viya also contains additional procedures that take advantage of the open, distributed environment.

1.9   Road Plan for the Book

The book covers many of the widely used predictive methods using SAS Viya. The book comprises eight chapters.

This introductory chapter discussed the difference and needs for data analytics and data science and highlighted the difference between the data analyst and data scientist. We identified and described the different analytics types and concluded the chapter with a discussion of SAS Viya, the leading platform used in this Book.

Chapter 2 focuses on the different data types and measurements, big data, data partitioning, honest assessment, an overview of the data preparation process, and the steps we take to connect data in SAS Viya. We will explore the SAS VIYA platform. The data set is a random sample subset of a more extensive database (over 600,000 records) for 2018 South West Flights (SWF) out of Florida on the Bureau Transportation Statistics (BTS) website (Bureau of Transportation Statistics 2020).

Chapter 3 focuses on the early stages of data exploration and highlights data visualization. We will also introduce the different descriptive measures and demonstrate how to load and explore data in SAS Viya.

Chapter 4 concentrates on evaluating predictive performance. Several measures are used and explained with their interpretations of the outcomes for assessing predictive performance.

Chapter 5 explains and demonstrates the use of a decision tree and ensemble for developing predictive models. Decision trees can model to predict a categorical variable or create a regression model for continuous variables.

Chapter 6 presents regression models, the math behind them, their uses, and then demonstrate regression modeling in SAS Viya using two different data sets: a classification example (a client who will subscribe to a term deposit or nonsubscriber) from the bank data set and a numerical variable from the concrete data set.

Chapter 7 presents neural networks. Neural networks are parametric nonlinear regression models. We create an interactive neural network and then create a NN through a pipeline demonstrating the different parameters and options and then compare the models. We show NN using the concrete data set. Concrete is an important material in civil engineering. The concrete compressive strength is a highly nonlinear function of age and ingredients used as our predictor variable.

Chapter 8 examines model deployment. It explains the deployment phase with its several tasks: Model assessment, model comparison, and monitoring model performance over time and updating as necessary.

Appendix A includes information about SAS VIYA for Learners and the many different offerings and materials for academics.

Appendix B includes a list of some open data sources.

Appendix C includes the data dictionary for the Aviation data set.

References

Bichler M, Heinzl A, van der Aalst WMP (2017) Business analytics and data science: once again? Business and information. Syst Eng 59(2):77–79. https://doi.org/10.1007/s12599-016-0461-1

Bureau of Transportation Statistics (BTS) 2020 Airline On-Time Statistics and Delay Causes. https://www.transtats.bts.gov/OT_Delay/OT_DelayCause1.asp?20=E

Burnham K (2019, Aug 29) Data analytics vs. data science: a breakdown. https://www.northeastern.edu/graduate/blog/data-analytics-vs-data-science/. Accessed 20 July 2020

Chen H, Chiang RHL, Storey VC (2012) Business intelligence and analytics: from big data to big impact. MIS Q 36:1165–1188

Gartner (2018). https://www.gartner.com/doc/reprints?id=1-1XYUYQ3I&ct=191219&st=sb

Gartner (2020a) Magic quadrant for analytics and business intelligence platforms

Gartner (2020b) Magic quadrant for data science and machine learning platforms

Informs (2014, Dec 18) Defining analytics: a conceptual framework. https://www.informs.org/ORMS-Today/Public-Articles/June-Volume-43-Number-3/Defining-analytics-a-conceptual-framework

Kiron D, Shockley R (2011) Creating business value with analytics. MIT Sloan Manag Rev 53(1):57

Maire S, Spafford C (2017) The data science revolution that's transforming aviation. https://www.forbes.com/sites/oliverwyman/2017/06/16/the-data-science-revolution-transforming-aviation/

Meulen R, McCall T (2018) Gartner survey shows organizations are slow to advance in data and analytics. Gartner

NAP Report (2014) Training students to extract value from big data: summary of a workshop. The National Academies Press. ISBN 978-0-309-31437-4. https://doi.org/10.17226/18981. http://nap.edu/18981

Provost F, Fawcett T (2013) Data science and its relationship to big data and data-driven decision making. Big Data 1(1):51–59

Richardson J, Schlegel K, Sallam R, Kronz A, Sun J (2021). Magic Quadrant for Analytics and Business Intelligence Platforms. https://www.gartner.com/doc/reprints?id=1-1YOXON7Q&ct=200330&st=sb

Schabenberger O (2020, Feb 24) Digital transformation projects don't fail because of a shortage of 'tech'. https://thenextweb.com/podium/2020/02/24/digital-transformation-projects-dont-fail-because-of-a-shortage-of-tech/

Shappell S (2006) Human error and commercial aviation accidents: a comprehensive, fine-grained analysis using HFACS. www.faa.gov/library/reports/medical/oamtechreports/index.cfm

Chapter 2
Data Types Structure and Data Preparation Process

Learning Objectives

- Identify the different data types and measurements
- Identify the need for data and data sources
- Explain data partitioning and honest assessment
- Identify the necessity of data preparation and curation
- Identify the data preparation process
- Explore SAS VIYA platform and import data

2.1 Data Types and Measurements

2.1.1 Taxonomy Data

Data are measurements or observations that are collected as a source of information. There are variety of different kinds of data and different ways to represent data.

At the highest abstract level, data may be classified as structured or semi-structured/unstructured.

Structured data are categorical (qualitative) data and numeric (quantitative) data and have the following characteristics:

1. **Qualitative or categorical data** can only be written in words; for example, the color of cars (red, green, black, etc.). Categorical data are grouped into categories that are based on qualitative characteristics and can take numerical values that do not have any mathematical meanings. Categorical data, in turn, are divided into nominal data and ordinal data.
2. **Quantitative data** can be written in numbers; for example, the cars' cost (1000, 2300, 4000, etc.). Discrete data are whole numbers; for example, the number of vehicles in the high street (2, 3, 5, 10, etc.). Continuous data are decimals, for

© The Author(s), under exclusive license to Springer Nature
Switzerland AG 2022
L. Halawi et al., *Harnessing the Power of Analytics*,
https://doi.org/10.1007/978-3-030-89712-3_2

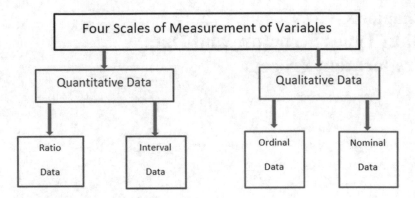

Fig. 2.1 Types of data and measurement scales

example, the weights of students in a class (45.56, 45.99, …., 52.35). Numerical data represent integers or discrete (only whole numbers) with interval scales or fractional numbers (real) or continuous with ratio scale, as shown in Fig. 2.1.

Briefly, data are also described in terms of their level of measurement. **Ordinal** scales provide information about the order of choices, and **nominal** variables are used to label or name a series of values. **Ratio** scales give a wealth of possibilities, and **interval** data are measured on an **interval** scale.

The categories of data examples are represented in Table 2.1.

On the other hand, unstructured/semi-structured data are composed of various textual, images, videos, voice, and web content. This variable needs to be converted to categorical or numeric representation before it is processed via analytics techniques.

It is important to note that some predictive techniques, data mining algorithms, and machine learning algorithms are selective regarding the type of data they can handle.

2.1.2 Big Data

Interest in Big Data has recently climbed dramatically. Big Data, in simple terms, is a set of data that cannot be managed, processed, or analyzed with commonly available applications in a reasonable time. The challenge presented with Big Data is characterized by the four Vs.: Volume, Velocity, Variety, and Veracity. Four characteristics characterize big data:

- **Volume**: due to technological advances, large volumes of data are collected, yet how to create value from data that is deemed to be relevant remains a challenge.
- **Velocity**: refers to the speed of processing data to meet the challenges.

Table 2.1 Data categories examples

	Categorial	Numerical	Categorial	Categorial	Numerical	Numerical	Categorial
	Nominal	Ratio	Nominal	Ordinal	Ratio	Interval	Ordinal
Name	Sex	Age	Marital status	Educational level	Wage annual growth rate	Weight/kg	How do you feel on Monday?
Allen	M	34	M	3	1%	[50,60]	Ok
Ana	F	29	M	5	4%	[60,70]	Unhappy
Craig	M	41	S	5	4%	[70,80]	Happy
David	M	32	S	6	5%	[50,60]	Ok
Daniel	M	29	M	7	6%	[60,70]	Very unhappy
Allen	M	25	S	3	1%	[70,80]	Happy
Liz	F	52	S	4	3%	[80,90]	Very happy

- **Variety**: data today come in all formats ranging from text documents, multimedia, sensor captured data, and biometrics. Some of these data are unstructured or semi-structured.
- **Veracity** is a term coined by IBM that refers to conformity to facts, including accuracy, quality, and usefulness.

2.2 The Need for Data and Data Sources

You want to solve a problem and make an informed decision. You have to connect with various data sources before knowing the best way to summarize, analyze, and visualize this data. All data measure something, or all data are variables with a measurement unit that can be coded and attached to the variable. There are two sources for data as described below.

- **Primary data** are collected by researchers from first-hand resources. Methods used are surveys, interviews, or experiments.
- **Secondary data** are gathered from studies, surveys, or experiments that have been run by other people or for additional research—for example, researching the Internet, newspaper articles, and company reports.

2.3 Data Sources

From the beginning of history, organizations have always needed to collect and store data. Data differences can go well beyond name-formatting on an Excel spreadsheet in the real world. Related data may be stored on entirely different applications and in various storage media. Data may be missing or wholly corrupted too.

With analytics projects, you are not conducting any transactions that create new data; you merely use data that already exists in the system. That means you have to work with the user groups to determine what data they need to do their jobs well, what information they would like to have to do their jobs better, and how best the analytics tools can deliver that data.

Data collection is expensive and time-consuming. In some cases, you will be lucky enough to have existing datasets available to support your analysis. You may have datasets from previous investigations, access to data providers, or curated datasets from your organization. In many cases, however, you will not have access to the data you require to support your analysis, and you will have to find alternate mechanisms. The data quality requirements will differ based on the problem you are trying to solve.

As a data analyst or data scientist, it is essential to be aware of your decisions' implications. You need to choose the appropriate tools and methods to deal with sourcing and supplying data.

Appendix B contains a list of data sources to use in SAS Viya operations or similar software.

2.4   Data Partitioning and Honest Assessment

To provide the required data for the analysis, we must consider the analysis period, estimation methods, variables, and data partition to generate learning/testing data and random samples. When deploying the model, there is usually a gap between using the model and the data used. In this case, the target variables typically refer to a different time from the input variables.

The strategy for choosing model complexity in data preparation is to use honest assessment. We split the data into training, validation, and testing dataset (optional).

Usually, a series of models are constructed, and the models increase in their complexity. The idea behind creating a series of models is that some will be too simple (underfit), and others will be too complex (overfit).

A portion is used for fitting the model, that is, the training dataset. The remaining data are separated for empirical validation. The validation dataset is used for monitoring and tuning the model to improve its generalization. The tuning process usually involves selecting among models of different types and complexities. The tuning process optimizes the chosen model on the validation data and finds the sweet spot between underfitting and being overfit.

The *test dataset* is optional for the model building process, but some industries might require it as a source of unbiased model performance. The test data give honest, fair estimates of the model. The test dataset provides one final measure of how the model performs on new data before the model is put into production. It assesses only the last model's performance on unused data. However, it might also be used to select the best model by scoring the champion model based on the training and the validation data.

2.5   The Necessity of Data Preparation and Curation

In today's world, we are continually generating, collecting, and retrieving data. Almost every action we take leads to a data point. Although this alone might seem like a lot of data, it doesn't even take into consideration the data generated from smart devices such as smart appliances, connected cars, and sensor networks, or operational data from organizations such as airlines, the stock market, university systems, research institutions, and other companies. The Internet of things (IoT) involves collecting and tracking data and applying analytics to the data *as it is gathered* to manage those things intelligently. Beyond the sheer amount of data, the speed at which data is produced and collected has drastically increased in the past 10 years. As we accumulate more and more data, organizations must learn how to leverage their data to make smart operational decisions.

Enterprise data include remarkable opportunities and considerable challenges. Data promises to alter businesses by generating new markets, raising revenue, and pushing new opportunities. However, we cannot fit and assess machine learning algorithms on raw data. Instead, this data needs to be cleansed, standardized, or transformed to meet machine learning algorithms' requisites. Additionally, we need to select an interpretation for the data that best reveals the prediction problem's indefinite structure to the learning algorithms to find the best fitting model.

The process of finding, exploring, preparing data for analytics and building models, structuring, updating, and even archiving data is called data curation. This is a critical step and a fundamental precursory step to building a model. We often use historical information, and a target variable must be present in the training data set.

Decision-making is data-driven. Data in all forms are recognized as strategic assets. The ability to collect and, at times, extract useful knowledge from data continues to be increasingly critical in today's highly competitive world. For the first step, you need to set your objectives and define the business problem or challenge clearly. Determining what data to use depends on the project's scope and the question you are trying to answer. Once you have useful data, it now needs to be prepared for analysis. Data preparation is a crucial step that is often overlooked. Data in the original form are not usually ready to be used in analytics tasks.

2.6   Data Preparation Process

For most analytics users, more than 70% of their time is spent in this process, or they are waiting for someone else to prepare the data for them. The data's transformation and aggregation are necessary to get relevant, reliable, and repeatable results out of the analyses. The type of collection has a significant impact on the final results.

The purpose of data preparation or commonly known as data preprocessing is to take the data identified, address the well-defined business task or problem, and prepare it for analysis. Understanding the project, how the various stakeholders use the

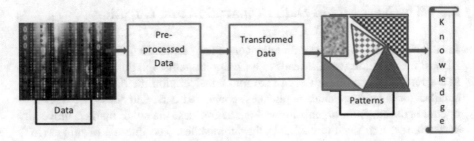

Fig. 2.2 Data selection phase

Fig. 2.3 Data preparation process

results, who will be affected by the results, is a precursor for this process. Figure 2.2 highlights the data selection phase.

Data preparation is not strictly about correctly transforming and cleaning existing data. It also includes a good understanding of the features that need to be considered and ensuring that the data at hand are appropriate in the first place.

The analytical data preparation process comprises the following steps, as shown in Fig. 2.3: Find, Assemble, Explore, Transform, and Improve.

1. Finding the dataset for the analysis involves sampling from a large database and data from different databases, internal and external, and other sources.
2. Assemble *Data blending* integrates data from disparate data sources into a form usable by analytical processes, models, and reports. Data preparation always starts with accessing data. The first step in the data preparation methodology, Manage Data, is to ingest data into the Software environment.
3. Explore data to verify that they are in reasonable condition and consistent in definitions, measurements, and periods. Are the values in a suitable range, and any missing data and any outliers?
Chapter 3 will cover data exploration and visualization.
4. *Data* preparation or wrangling is the process of transforming and cleansing dirty data into the desired format that is "fit for purpose." Data preparation includes data selection, preprocessing of data, data sampling, and data conversion; this is the most resource-consuming step. Figure 2.4 summarizes the data preprocessing steps.

1. Data selection or identification and data extraction, including noise removal, time sequences identification, and missing variables strategies.
Missing values are essential to be handled correctly to analyze data successfully.
Missing data is a common problem. To address this problem, we need to know

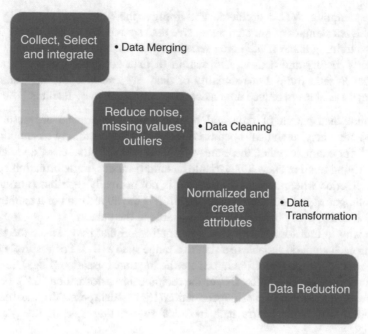

Fig. 2.4 Data preprocessing steps

the difference between "real missing data" and "not stored data." A realistic example of missing data is a dataset of individuals who may have missing values for existing medical conditions, COVID-19 infection status, whether they have taken an anti-malaria drug and the success or death rate post-anti-malaria medicine. As we can see from these examples, real missing data occurs in datasets generated from situations where that information is lacking. If real missing data is detected, we can use some imputation methods devised for restoring missing values in the data based on additional entries in the database. The choice of the imputation method will impact the performance of the machine learning technique. Also, the imputation method will differ based on the business context and any other existing information. Some imputation methods can be simple methods such as using the mean or median or more advanced methods such as a regression technique. On the other hand, when we have missing data resulting from "not stored information," we may replace this missed data with any number or value or the business meaning of that data. Alternatively, If the number of cases missing values is small, we may omit those values from the analysis.

Outliers are the significant values in the dataset. They may be usual occurrences or may be due to processing errors. We can opt to dismiss these outliers from the analysis. It is recommended that you compare the report with and without outliers before you make your decision.

2. Data sampling. We can accelerate and improve the building of models by taking intelligent samples of our comprehensive data source. We usually divide the data into a training dataset, validation set, and optionally test dataset.
3. Quality testing and data transformation or data conversion into an analytical model to reduce the dimensionality of data. We can also extract features and change variables to reduce dimensionality or generate new features.

Normal distribution of data and homogeneity of variance are assumptions required for many analytical operations. Suppose the normality assumptions are violated, for example, when there are variable outliers. In that case, a variable has an underlying trend or skew. If a variable has multi-modes, transformation becomes a good practice when the data distribution is not normally distributed or does not have homogeneity of variance. Transformation is the application of a mathematical operation across every observation of a variable in the dataset.

One way to transform data includes normalization that may reduce the range of values in each numeric variable to a standard range such as 0–1. For positively skewed data, the most common transformations are the log transformation (Logx), the square root transformation (SquareRootX), or the reciprocal transformation (1/X). For negatively skewed data, reverse score transformation (Highest data point-X) is most common.

Another way to transform data includes aggregating data or the process of discretizing.

Also, we can construct a new variable using a wide range of mathematical functions such as multiplication and log transformation.

Dimension reduction is a technique to reduce the number of input variables. *Principal component analysis* (PCA) is a commonly useful method for dimension reduction, especially when the number of variables is large. When you notice that you have subsets of measured measurements on the same scale and are highly correlated, PCA is beneficial in this case. PCA provides fewer variables that are a weighted linear combination of the original variables and thus retain most of the full dataset (Table 2.2). It is used for numerical variables, categorical variables, correspondence analysis, and other more suitable methods.

Remark: Data preprocessing is a tedious and time-consuming process meant to improve data selection quality. This is a vital step often overlooked. Datasets may be enormous. For the most part, the original data collected may not be ready to be used. The preprocessing involves enriching the data with external data. This step consists of a sub-step where we remove any noise in the data. Noise may be

Table 2.2 Data preprocessing dimension reduction

Dimension reduction	Descriptions
Feature extraction	Transforms variables into lower dimensions; this is done by using PCA and other techniques
Variable clustering	This node divides numeric data into disjoint clusters or groups and chooses a variable to represent each cluster, thus reducing redundancy and decreasing multicollinearity
Variable selection	Uses different supervised and unsupervised methods to determine which variable has the most significant impact on the model

Fig. 2.5 SAS drive

redundant data, missing variables, and outliers. The best guidance is to be consistent with practices from your discipline. This process can take time and trials until you achieve data that is appropriate for analysis.

2.7   Exploring SAS VIYA Plaftorm and Import Data

SAS Viya is an architecture that helps organizations access, explore, transform, investigate, and even govern their data. SAS Viya enables different users to cooperate and share insights.

When you first open SAS VIYA, you will be on this page (Fig. 2.5). Spend the time exploring first your application workspace, the SAS Drive.

2.8   Understanding the SAS VIYA Interface

SAS Drive is always available from the Applications menu in the upper left.

1. Applications menu.

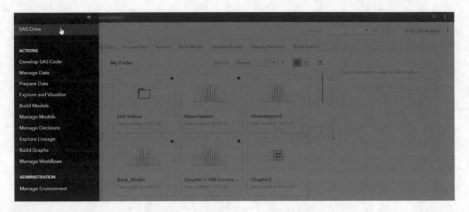

Fig. 2.6 Application menu

If you select from the top far left of the screen right by the SAS drive—Organize and share content. The menu opens with a possible list of actions from developing SAS code to preparing data, among other options. The availability of the features presented in Fig. 2.6 in the SAS drive depends on the applications installed based on your license.

2. New item button.

New ▼   icon. If you open this icon, a menu appears with three possible actions: using a link with the URL and specifying the location of the relationship; creating a shortcut for a new project or Folder with a designated area in the SAS CAS library; and lastly, uploading with two upload options to upload files from your computer to my Folder or to upload a folder from your computer to my folders.

We will demonstrate the use of the upload in the last section of this chapter.

3. Folders and Filter.
The Tiles area contains tiles with recent content, favorites, links, and any content tiles that have been added by the user or an administrator. There is also the filter.

4. The Search feature allows an advanced search by item name, by type such as a data model, a folder, and created and modified by user name from the list of members who have access to the system.

5. Undo and Redo. Click and hold on either icon to display a list of actions.

At any point in time, you can select   to do and undo a given selection.

The quick access area at the top far right of the interface allows you to store the most frequently used items and pin an item to the Quick Access area. By default, the SAS videos link is pinned, and when you right-click it, you may unpin from the quick access area.

6. Menu. Create links or shortcuts, manage tabs, upload content.
   The tabs are used for arranging and sharing content. The All tab displays all the items that you have access to. The projects tab, for example, allows you to group content and data, control membership, and view activity.
7. In this area, the recent items, notifications, help, settings, and sign-out options are available.
8. This is the information pane area, once we have a selected folder, model, summary information, and comments will be displayed.
9. This is the canvas area where all the projects you create will be displayed.

2.9   Loading Data into SAS VIYA

Model Studio within SAS Viya is a central web-based application that supports the analytics cycle. All imported data is available on the CAS server through caslibs. Caslibs are an in-memory location with data tables and access control lists; it also contains other data source information. They may be predefined, manually added, or personally set.

There are different ways to load data into the SAS VIYA environment. With SAS Viya, you can import different types of data saved as spreadsheets, text, or SAS datasets; you can connect to a database such as Oracle or Hadoop. You may also import social media data once authenticated by the source.

The dataset for the example is a random sample subset of a larger database (over 600,000 records) on the Bureau of Transportation Statistics (BTS) website (Bureau of Transportation Statistics 2020).

This more extensive BTS database was created for the Research and Innovative Technology Administration (RITA). RITA is a unit of the US Department of Transportation (USDOT). USDOT was created in 2005 to advance transportation science, technology, and analysis and improve transportation research coordination within the department and throughout the transportation community. RITA's basic functions offer transportation statistics for use in decision-making, research, and education. The sample dataset includes variables for a partial month during a year for one carrier's flights out of one state. The variables include identifying information about flights and other characteristics that can affect on-time performance for each flight.

The dataset consists of 44 columns (variables) and 1161 rows. Some of the variables include Year (numeric), Month (categoric), Day of the Month (numeric), Day of the Year (numeric), Flight Date (numeric), Carrier (categorical), Tail Number (numeric), Origin (categorical), Destination (categorical) among other variables. In Chap. 3, we will explore this dataset.

For this demonstration, we will start by selecting the Prepare Data options from the menu.

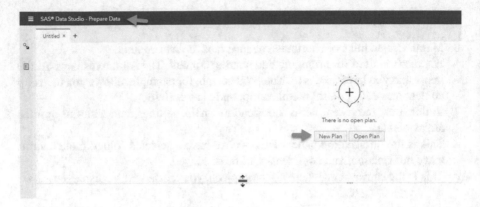

Fig. 2.7 New plan

Fig. 2.8 Import local data file

You can create a new plan then import your local data; in this case, the excel file is saved on your computer; this is why we are using the import tab. Once the import is complete, a confirmation for the successful import will be displayed. You are ready to transform your data, create reports, explore and visualize your data, and build models (Refer to Figs. 2.7, 2.8, and 2.9).

Go to the prepare data tab under the SAS Menu and select the new plan (Fig. 2.7).

Select the import option and local the file that you want to import into the SAS CAS memory; in our case, it is a local file saved in our computer, as presented in Fig. 2.8.

Select the import item option as shown in Fig. 2.9.

Once the import is complete, a confirmation for the successful import will be displayed (Fig. 2.9). If you select Ok, you will be taken to the work area.

Another way to load data into the system is through SAS Visual Analytics, the Explore and Visualize tab, shown in Fig. 2.10. We can start with data, and this will only import the data into the memory.

Fig. 2.9   Confirmation of import

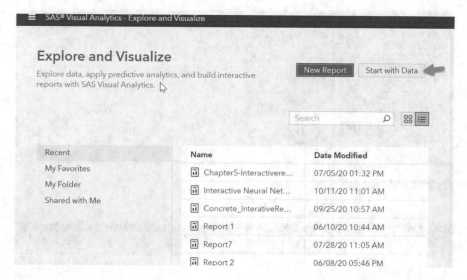

Fig. 2.10   Start with data in SAS visual

We select the import and local our file from the local computer and select import as presented in Fig. 2.11. We typically follow the same path presented in the previous demonstration.

The 🗲 indicates that the file is loaded into the memory and is ready for use (Fig. 2.12).

Once you press Ok, the report area is created, and you can start the explore and visualize the process as presented in Fig. 2.13.

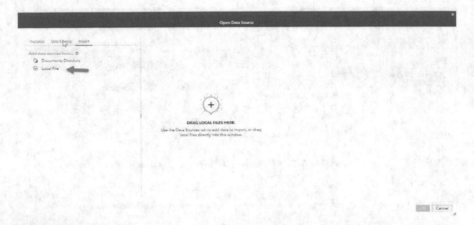

Fig. 2.11 Import data

Fig. 2.12 Uploaded data

Fig. 2.13 Report area

Reference

Bureau of Transportation Statistics (2020) https://www.transtats.bts.gov/databases.asp?Z1qr_VQ=E&Z1qr_Qr5p=N8vn6v10&f7owrp6_VQF=D

Chapter 3
Data Exploration and Data Visualization

Learning Objectives
- Explore data for modeling and exploration
- Use and interpret descriptive statistics
- Identify and demonstrate the different types of visualization techniques
- Recognize the value of data visualization
- Explore and Visualize data with SAS Viya

3.1 Aviation Case

With data, airlines can make informed decisions and act faster. These decisions transpire into particular advantages for improving the customer experience as problems can be anticipated and possibly mitigated before becoming critical incidents.

In 2019, the average cost of aircraft block (taxi plus airborne) time for U.S. passenger airlines was $74.24 per minute. Delayed aircraft is estimated to have cost the airlines several billion dollars in additional expense. Delays also drive the need for extra gates and ground personnel and impose costs on airline customers (including shippers) in the form of lost productivity, wages, and goodwill. Assuming $47 per hour* as the average value of a passenger's time, flight delays are estimated to have cost air travelers billions of dollars. FAA/Nextor estimated the annual costs of delays (the direct cost to airlines and passengers, lost demand, and indirect costs) in 2018 to be $28 billion.

Given these costs, aviation organizations can use this dataset to analyze operations to inform variables' impacts on events, causes of events, and most importantly, indicate where changes can be made to avoid events that add costs to the airline's bottom line. In our example, an aviation executive can ask an analyst to analyze all flights' on-time performance out of a particular state. Since time is money, the

© The Author(s), under exclusive license to Springer Nature
Switzerland AG 2022
L. Halawi et al., *Harnessing the Power of Analytics*,
https://doi.org/10.1007/978-3-030-89712-3_3

analyst can focus on trends in delays, make predictions, and propose enhancing operational solutions with precision and agility.

The dataset, for example, is a random sample subset of a larger database on the Bureau of Transportation Statistics (BTS) website (Bureau of Transportation Statistics 2018). This dataset is a sample of all flights of one airline departing during 1 month in Florida. This larger BTS database was created for the Research and Innovative Technology Administration (RITA). RITA is a unit of the US Department of Transportation (USDOT). USDOT was created in 2005 to advance transportation science, technology, and analysis and improve the coordination of transportation research within the department and throughout the transportation community.

This sample dataset includes variables for a partial month (2018) for one carrier's flights out of one state. The variables include identifying information about flights and other characteristics that can affect on-time performance for each flight. It consists of features categorized as follows (Appendix C).

Information About Flight: year, Day, Day of Month = Day in the month, Day of Week = Day of the week starting with Sunday = 1, tail number, carrier, flight number.

Information About Origin: origin, origin city, origin state.

Information About Destination: destination, destination city, destination state.

Information About Departure: CRS DepTime: Scheduled Departure Time (expressed in military time 2400), DepTime: Actual departure time (expressed in military time 2400), DepDelay: Departure Time − CRS DepTime (in minutes), DepDelayMinutes: if DepDelay >0, = DepDelay; otherwise zero (in minutes), DepDel15 = if DepDelay >15, = 1; otherwise zero (ordinal data), DepartureDelayGroups = if DepDelay <15 = 0, if DepDelay between 15 and 30 = 1, if DepDelay between 31 and 45 = 2, etc. (ordinal data), TaxiOut = minutes taxing from DepTime to Wheels off (expressed in minutes).

Information about Flight: distance, distance group, ActualElapsedTime, CRSElapsedTime, ArrTimeBlk.

Information about arrival: WheelsOn, TaxiIn, CRSArrTime = Schedule Arrival Time, ArrTime: Actual arrival time, ArrDelay: Departure Time, ArrDelayMinutes, ArrDel15, ArrivalDelayGroups, WheelsOn, TaxiIn.

Information about cancelation and unusual events: CarrierDelay: (expressed in minutes), Weather Delay: (expressed in minutes), NAS Delay: (expressed in minutes), SecurityDelay: (expressed in minutes), LateAircraftDelay: (expressed in minutes), canceled, diverted.

Questions that we investigate in this Chapter in the SAS VIYA demonstration section:

1. What are the percentage delays by origin city?
2. What are the arrival delays per flight date and the five numbers summary?
3. Which city has the highest frequencies of delays?
4. Which destination city has the highest traveled distance and arrival delays?
5. What is the total breakdown of the arrival delays with the days of the week and the Original city
6. Can we detect any pattern within the frequencies of flights per Day of the month, Day of the week?

3.2 Investigate Phases of Data and Data Exploration

Who does not want to be able to answer questions and solve problems? The educated way to answer questions and solve problems is to conduct a systematic research process. That research process must involve data. Whatever discipline the research takes place in requires collecting and exploring data to generate explanations and predictions about a hypothesis.

Data is made of variables that, by definition, are measurements of things. Research scientists decide what they want to measure depending upon what they want to study. Values such as flight time, departure delay, time of arrival, and airport destination, to name a few, are all examples of data that measures something. A researcher can generate new ideas, probabilities, correlations, and mathematical models by exploring data and datasets. Research scientists can use data to observe phenomena about our questions without directly impacting any outcomes. This can only be looking at the time in minutes a student spends in a statistics class and the resulting test score on a final. Research scientists can also use data to manipulate and generate additional data to come to informed conclusions. Using the classroom scenario, research scientists can partition data into groups to see if groups scored differently on a final from each other, add in other behavioral data about resources they used, etc., to generate different theories about how groups score on a final. The exploration possibilities are endless, given the data and software to analyze.

We may come across many issues during data collection that may impact our estimation. Among the most commonly cited issues, data may not be available and may require a fee to access it. Data may be inaccurate or may be insecure. The concept of GIGO—garbage in, garbage out, is ever-present in any estimations, modeling, and summary of data. Incorrect data used in the analysis to reach conclusions will produce faulty output and, thus, findings from the data.

Exploratory data analysis (EDA) is the first task that a data set goes through. EDA helps us to understand our data to help prepare for other tasks. As presented in Fig. 3.1, exploring data encompasses many phases or steps. We need to start looking at the size of the data, its shape (symmetric, left-skewed, right-skewed, curvilinear, or others), how many columns we have, how many rows, content type, how many characters, and the type of the variable (numerical, qualitative, or text data).

Exploration is the primary step in data analysis. Users explore the dataset in an unstructured way to reveal initial patterns, characteristics, identify the missing values and outliers and points of interest. Data exploration can utilize manual methods and automated tools such as data visualizations, charts, quantitative and initial reports. It suggests the next logical steps, questions, or research areas for your project.

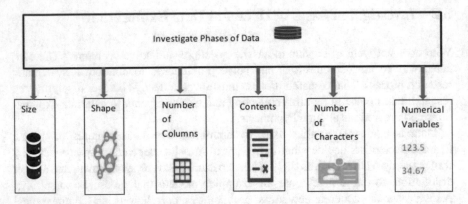

Fig. 3.1 Investigate phases of data

3.3 Putting Descriptive Measures Together

If someone asked you to summarize a variable, you would describe it. Hence the name is Descriptive Statistics. Descriptive Statistics provide essential distribution characteristics about the data involved in a dataset or study and form the basis of more advanced quantitative analysis and inferential statistics. Descriptive statistics mainly describes the data in a sample or Population in a simple summary format.

For example, if you are a frequent traveler, you may want to know an air carrier's on-time status on a particular route as you make your arrangements. The US Bureau of Transportation Statistics reports data on airline flights. For instance, consider a simple number used to summarize how well an air carrier performs on a specific route concerning arriving on time: the on-time arrival percentage. This single number is simply the number of on-time arrival flights for a carrier on the route divided by the number of flights conducted within a specific period. A carrier reporting 0.75 is arriving late one time in every four flights. One reporting 0.50 is arriving on time two times in four. This single number describes many discrete events (Bureau of Transportation Statistics n.d.).

Now expand our example to a large number of observations in a dataset. A dataset is a group of scores or measures of variables. Researchers beginning to understand what the data reflects often want to know about the scores' descriptive statistics through central tendency measures, measures of Variation, and distribution shape. There are three standard measures of central tendency: mean, median, mode (defined subsequently). Standard measures of dispersion or variability about the variables are the range, variance, and standard deviation (described previously). Researchers describe the Sample data or a visualization of the data between central tendency and variability measures. And finally, researchers would be keen to know whether the data is skewed or symmetric, which would indicate if there are extreme observations of a variable.

3.4 Descriptive Statistics

Descriptive statistics are appropriate when the research asks: What percentage of X, Y, and Z participants? How long have X, Y, and Z been in a certain group? What observations we can make about participants, the count, the mean amount spend per customer.

In the descriptive approach, we explore the data by inspecting the summary statistics such as Measures of Central Tendency, Measures of Variation, and Distribution Shapes (Fig. 3.2). Figure 3.2 shows the descriptive statistics measures and their uses.

The central tendency is the extent to which the numerical values group around a primary value. The variation or dispersion is scattering away from the central value. The shape of the distribution is the pattern presented from the lowest to the highest value.

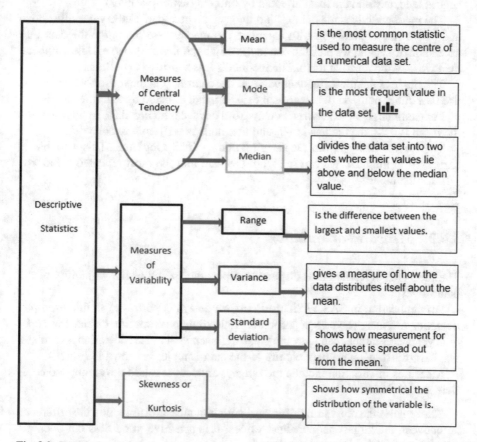

Fig. 3.2 Descriptive statistics measures

3.4.1   Measures of Central Tendency

The mean, median, and mode are all valid central tendency measures, but some central tendency measures become more appropriate to use than others under different conditions. In the following sections, we will look at the mean, mode, and median and learn how to calculate them and under what conditions they are most appropriate to be used.

The mean or (average) is the most common statistic used to measure the center of a numerical dataset, summarizes a large amount of data into a single value, and can be used with different data types. The mean is the sum of all the values divided by the number of values in the dataset. If we have n values in the dataset, and the values are $x1, x2, \ldots xn$. The sample mean formula is denoted by $x^- = \sum fi\, xi/n$ where n is the number of values or the sample size, fi is the frequency of xi, and n is the number of terms in Sample $1 \leq I \leq n$. The mean is the most common measure of central tendency and the most affected by outliers (extreme values).

The median divides the dataset into two sets, where half of the values lie above and below the median value. To calculate the median, we rearrange the data into ascendant order, and we pick the one in the middle; if there are two middle numbers, then we take the mean of them. The median is less sensitive to outliers.

The mode is the most frequent value in the dataset; it is the most popular option used for either numeric or categorical data. It is not affected by extreme values.

For example, if we take arrival delay from our dataset (Fig. 3.3), we can see that the mean is 0.43, the median is −7, and the mode is −10 (min in delays).

The arrival delay is the Departure Time − CRS DepTime. This variable is expressed in minutes. We can see from the data that the most repeated value was −10 for the mode.

3.4.2   Measures of Variation

The measures of Variation or dispersion report the spread or the variability of the data values.

An application in SAS VIYA using arrival delays is portrayed in this example. Selecting arrivals delays for a particular flight route on a specific carrier for a specific year, a researcher can view the measures such as standard deviation, skewness, and Kurtosis to get a primary picture of the data Sample, as shown in Fig. 3.4.

Measures of variation include the range, variance, standard deviation, and coefficient of Variation.

- The range is the most straightforward variation measure, reflecting the difference between the largest and smallest values. It is sensitive to outliers but does not account for how data are distributed.
- The variance (σ^2 for population or S^2 for Sample) is the average of the values' squared deviations from the mean. The variance is expressed in the squared

Fig. 3.3 Descriptive metrics–arrival delay

Fig. 3.4 Measures details arrival delays generated by SAS VISUAL VIYA

Standard Deviation:	28.14
Standard Error:	0.83
Variance:	791.74
Distinct Count:	133
Number Missing:	13
Total Observations:	1,150
Skewness:	2.8890
Kurtosis:	11.4159
Coefficient of Variation:	6,497.6916
Uncorrected Sum of Squares:	909,922.00
Corrected Sum of Squares:	909,706.34
T-statistic (for Average=0):	0.5219
P-value (for T-statistic):	0.6018

units. The variance for arrival delays in our prior example is equal to 791.74 (Fig. 3.4).

- The standard deviation is the most commonly used measure of Variation; it shows the Variation about the mean and expressed in the same units as the mean and the original data. It is the square root of the variance, and its symbol is σ (the Greek letter sigma) for the Population and S for the sample. The standard deviation for arrival delays in our prior example is equal to 28.14 (Fig. 3.4).

- The coefficient of variation (**CV**) measures the relative variation and is always in percentage form. It shows the variation relative to the mean and may also be utilized when comparing the variability of two or more sets of data measured in different units. $CV = (S/x^-)* 100\%$. Where x^- is the Sample mean, and S is the sample variance. The coefficient of the variable for arrival delays in our example is equal to 6497.6916 (Fig. 3.4).

Measures of variation application examples are shown accordingly in Fig. 3.4.

Remark: It is essential to note the difference in symbols and denominators when drawing data from a population instead of a sample.

N is the population size, n is the sample size.

When you have "N" data values are collected from the Population, we need to divide by N when calculating variance. When you have "n" data values obtained from the Sample, we need to divide by $n - 1$ when calculating variance.

The more data are spread out, the higher the values of the measures of dispersion. If all values are the same, the measures of dispersion are equal to zero. It is also important to note that none of these measures are ever negative.

3.4.3 Distribution Shapes

The shape of a distribution depicts how data are distributed. We can look for skewness in the data or the Kurtosis.

The skewness measures the extent to which the data values are not symmetrical; they may be left-skewed where we can notice that the mean is less than the median value or right-skewed where the median is less than the mean if not symmetric and in this case, the mean is equal to the median. Our skewness is 2.8890 for arrival delays, as presented in Fig. 3.4.

The frequency distribution presented in Fig. 3.5 combined with the results of the measures of variations from Fig. 3.4 can tell us visually where the mean, median, and mode are. We can see the median in the red line vertically with (-7), and the mode is -10, and we can mostly see the distribution shape of the data.

Kurtosis measures how sharp the plot of data or curve rises, approaching the center of the distribution. We notice a leptokurtic or more pointed peak with Kurtosis >0, a bell-shaped or Mesokurtic peak with Kurtosis = 0, and flatter than the bell shape peak or Platykurtic peak with Kurtosis <0. Our Kurtosis for the arrival delay variable is positive at 11.4959, as presented in Fig. 3.4.

The five-number descriptive summary helps detail the center spread and shape of data as well.

The five-number summary consists of the smallest number, First Quartile Q1, Median or Second Quartile Q2, Third Quartile Q3, and the largest number.

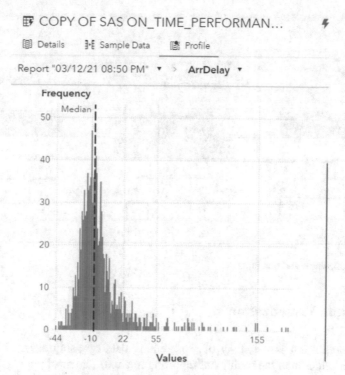

Fig. 3.5 Distribution type arrival delays

Table 3.1 Relationship between the five-number summary and distribution shapes

Left-skewed	Symmetric	Right-skewed
Median − $X_{smallest}$	Median − $X_{smallest}$	Median − $X_{smallest}$
>	≈	<
$X_{largest}$ − Median	$X_{largest}$ − Median	$X_{largest}$ − Median
Q_1 − $X_{smallest}$	Q_1 − $X_{smallest}$	Q_1 − $X_{smallest}$
>	≈	<
$X_{largest}$ − Q_3	$X_{largest}$ − Q_3	$X_{largest}$ − Q_3
Median − Q_1	Median − Q_1	Median − Q_1
>	≈	<
Q_3 − Median	Q_3 − Median	Q_3 − Median

Quartiles divide the data into quarters, each containing an equal number of data. They are useful when data is not symmetrically distributed. The lower quartile (Q_1) represents 25% of the data, the median (Q_2) represents 50%, and the upper quartile (Q_3) represents 75%. The interquartile range (IQR) is the difference between the upper (Q_3) and lower (Q_1) quartiles. We can notice a clear relationship between the five-number summary and distribution shapes in Table 3.1 and Fig. 3.6 below.

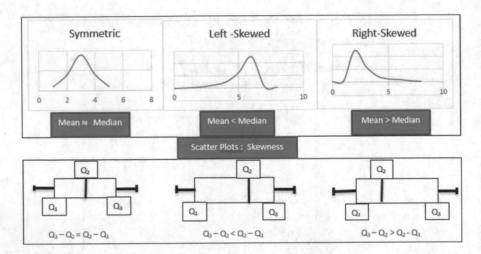

Fig. 3.6 Skewness and box plots

3.5   Data Visualization

Data visualization is the study of representing data by using a visual or artistic approach rather than the traditional reporting method. The most popular data visualization types are dashboards and infographics, which use charts, text, and images to communicate the data's message. Data visualization is closely related to information graphics.

With the recent popularity of social media and mobile apps, the amount of generated data is astounding. For this reason, many companies find that making sense of that data requires the use of some form of data visualization.

Data visualization is used in the data mining process's preprocessing phase when we clean data by locating incorrect values, missing values, and removing redundancies and duplications. These data visualization techniques are also useful for selecting variables for the predictive model and data derivation and data combining, a critical part of the data reduction process.

This chapter focuses on the exploration phase of data and graphical presentations, mainly using SAS VISUAL ANALYTICS VIYA. As we started the chapter, data exploration is the mandatory initial step in any analytics project.

3.5.1   Data Visualization Considerations

While data visualizations may be used for everybody, not necessarily only people in decision-making, we know to acknowledge that useful visualizations may come in various shapes and sizes.

Using visuals with your data to tell a story is the most widely used science, business, and education method to convey complex information quickly and with a minimum explanation. Knowing how to create visuals that are useful, concise, and elegantly tell the story behind the data can be one of the most valuable career skills to acquire. Figure 3.7 summarizes some rules for data visualizations.

When creating a visual, you need to consider the following elements:

- Your visual needs to speak to a specific audience
- You need to choose the Right Visual Type guided by the purpose of the message
- You need a Clear Visual to convey the message in simple ways
- Do you need to answer the questions you are asking? Does the Visual answer them?

According to Yarmuluk (2019), there are four critical traits of powerful data visualization are:

1. Graphical integrity (truthful and honest picture)
2. Design (striking a balance between eye-catching visuals and effectiveness)
3. Interactivity if possible
4. Color (use color to enhance the visuals and overall appeal).

Howard and Thompson (2015-2016) considers data visualization as a dynamic process and highlights seven characteristics of compelling visualizations. The first important trait is knowing the point you are trying to make and choosing the right graph. Do not overdo the formatting once you select the right figure, less is more. Also, use color when you need to pass a message while creating pointed titles that link to the printed text. Lastly, revise based on any feedback received.

McCandless sums up the traits of useful visualizations with these simple yet powerful traits—the story as a concept, the goal or function, the data/information, and the visual form of metaphor.

According to TechNews (2016), visualizations should be visually appealing, scalable, provide users with the correct information accessible, and allow rapid development and deployment.

Aside from the rules for creating visualizations, some characteristics make visualizations a powerful tool.

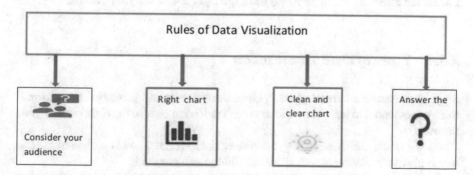

Fig. 3.7 Rules of data visualization

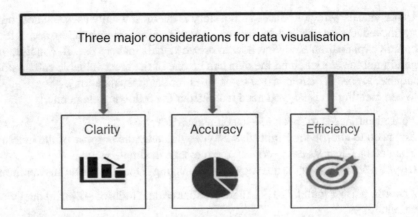

Fig. 3.8 Data visualization considerations

Three primary considerations, when creating the data visualizations, refer to Fig. 3.8.

1. Clarity: to ensure that the dataset is relevant and complete.
2. Accuracy: to ensure the right graph is representing the data.
3. Efficiency: to ensure efficient visualization techniques are used to highlight the data.

According to TechNews (2016), the three considerations are meaningful, desirable, and usable visualizations.

Many techniques and best practices are available for creating compelling visualizations in the market. Figure 3.9 outlines our approach to data visualizations along with the factors that we take into considerations.

An excellent data visualization should be visually appealing, offer accurate and correct information for the different variable types. It should be telling the story too.

Keep it simple, utilize the most critical information, don't overload your data visualization, and make sure that your data visualization requires little explanation. Labels are essential, along with choosing the right colors for your visuals.

3.5.2 Types of Data Visualization

Since data visualizations are used to empower the user and give accurate information, it is essential to learn the various visualizations to choose the right one for your purpose.

As we stated earlier, selecting the proper data visualization is a challenge. The key is always to choose attractive visuals that match our data.

We have many different content types to choose from, such as graphs with an x and y-axis, tables with columns and rows, a timeline highlighting a chronology of

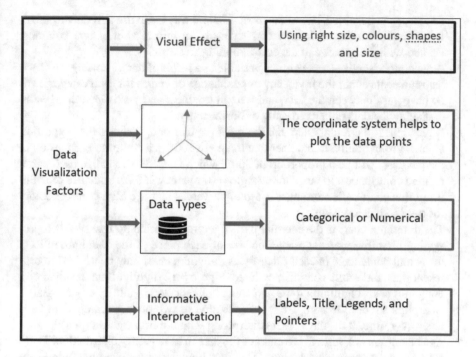

Fig. 3.9 Data visualization factors

events and changes, a flow chart with a sequential set of instructions. This map displays the big picture, among other content types.

Using visuals with your data to tell a story is the most widely used science, business, and education method to convey complex information quickly and with a minimum explanation. Knowing how to create useful, concise graphs and tell the story behind the data can be one of the most valuable career skills to acquire. We will review some of the most common graphs and tables used to summarize and explore data.

- A summary table counts the frequencies or percentages of items in a group of categories highlighting the differences between classes. Summary tables may be more useful if the central intention is to offer specifics for values lookup, exact values of the data; you need to briefly display information on the same topic that employs the same measurement unit.
- A Contingency Table Helps Organize Two or More Categorical Variables and study existing patterns across two or more categorical variables. We can create contingency tables based on the overall total, the percentage of row totals, or the column totals percentage.
- A frequency distribution is a summary table in which the data are arranged into numerically ordered classes. It is used for numerical variables and offers a quick visual interpretation of the data. You may determine some of the significant characteristics of where the information is concentrated or clustered. When a lot of

data needs to be sorted, one of the most efficient ways is to use a frequency table. It summarizes a set of data by classifying them into a table of two columns, labeled with the values and the corresponding frequencies.

- A bar chart visualizes a categorical variable as a series of bars. The length of each bar represents either the frequency or percentage of values for each category. We do have gaps between the bars, and we can use the x and y-axis interchangeably to display the data either vertically or horizontally.
- A pie chart is a circular chart and the most popular one. It shows the proportion of each group briefly. Remember that there are 360° in a circle, so each group in the pie chart will be a proportion of 360°, which is used to visualize each part's relative contribution to the whole. Angle = (frequency.360)/n. You need to avoid using this chart with more than six to eight values and use high-contrast colors for easy and clear viewing.
- The doughnut chart is the external part of a circle divided up into pieces representing the different categories. The size of each piece of the doughnut reflects the actual percentage in each category. A doughnut chart may be utilized to represent the data from a contingency table. Figure 3.12 highlights an example of a doughnut chart highlighting the percentage of canceled flights by city names.
- Box plots are used to compare the distribution of two or more categorical variables over time. It displays the median along with the first and last quartile.
- The histogram is a vertical bar chart of the data in a frequency distribution. There are no gaps between adjacent bars. The class boundaries (or class midpoints) are displayed on the horizontal axis. The vertical axis is either frequency, relative frequency, or percentage. The height of the bars denotes the frequency, relative frequency, or percentage. Histograms are typically used when the data is in groups (one measure).
- A-line chart consists of plotted data points on category and response axes and joined with straight line segments. We can use line charts to visualize a trend in the response data over category values. Figure 3.15 presents two examples of a line chart. The top part presents the destination city name's flight number, and the bottom line chart portrays the flight number versus the destination city name.
- Time-series plot is used to study patterns in the values of a numeric variable over time. Times series graph is a line graph that illustrates data points taken at regular intervals. On the x-axis, we plot the time increments, and on the y-axis, we plot the corresponding value. A scatter plot examines possible relationships between two numerical variables.
- Geo map presents the data in a location element on the map. Locations can be displayed as bubbles where the values of different measures are proportional to the size of the bubble, as color points where the color represents the measure of category on a discrete scale by using display rules, or as regions on a map where the color represents the measure on a continuous scale. A Geo map is used when data represent one geography and up to two measures. The geographic variable represents a special type of data variable where each item has a latitude and longitude value.

Note that Dashboards are a data visualization tool that allows all users to understand the analytics that matter to their business, department, or project.

3.6 Application Using SAS VIYA

Log in to SAS VIYA and Go to the Explore and Visualize Data Option as presented in Fig. 3.10.

We will illustrate how you can explore and visualize your data, answer the questions, and save our report to share with colleagues or our supervisor.

We did import the data in the previous Chapter. It is already in the CAS memory. Let us follow the following steps as presented in Fig. 3.11:

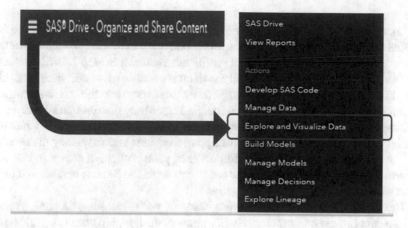

Fig. 3.10 SAS drive menu—explore and visualize data

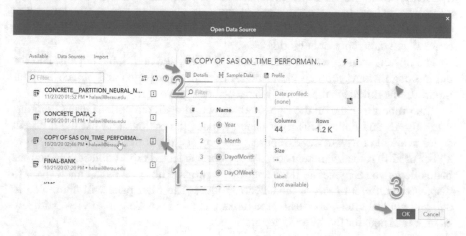

Fig. 3.11 Open the data source

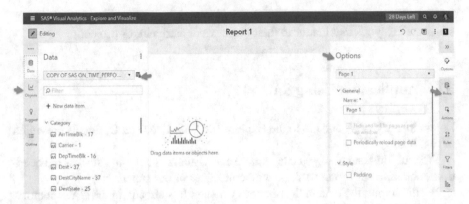

Fig. 3.12   Report area

Locate your data previously imported. **the** green arrow, if selected, will unload the data from memory, and the other one is for displaying additional options Review the details, sample data, and profile tab as shown in step 1. When selecting the details in step 2, you can see that the dataset consists of 44 columns (variables) and 1161 rows. We can also see different statistics for the selected variable (Descriptive Metrics, Pattern distribution, and frequency distribution). Lastly, press OK, as shown in step 3. The Data Explorer let us profile data to detect any inconsistencies and anomalies and review descriptive measures and frequency distribution.

The SAS Explore and Visualize Report Area pane is open, as shown in Fig. 3.12. It consists of three primary areas, the content pane on the left has buttons that toggle open the data, objects, and outline panes.

The data pane contains a list of data items. If you select the data pane, you will see the data broken into categories with numbers designating each categorical variable's different count. Measures represent numeric variables such as Airtime, ArrDelay, and aggregated measures such as the frequency percent. You can create additional aggregate measures. For example, as shown in Fig. 3.13, destination (dest) has 37 different categories or classes, and Origin State shows one, making sense as the data represents Florida only.

The objects icon offers the list of tables, graphs, and additional objects, and the outline page where you can view and work with your pages and objects in your report. In the middle is the canvas where you design your report.

The properties pane on the right enables you to open the options, roles, actions, rules, filters, and ranks. This right pane includes the options that list the different styles available for your page, reports, and objects; the Roles pane enables you to add or modify role assignments for the currently selected report object. The actions pane allows you to generate links and filter actions. The rules pane enables you to display the different rules of the selected object. The filter pane will allow you to add and control filters, and lastly, the ranks pane will enable you to view, add, and change rankings for the selected report object.

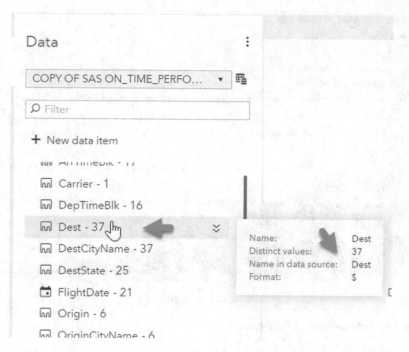

Fig. 3.13 Data pane

We can see the data broken down into categories and measures on the left pane. If you select 🖳 next to data, the measure details option displays all the summary statistics.

When you drag a data item from the data pane and drop it onto the canvas, SAS visual analytics will create an appropriate graph for the item. We can have different visualizations, and we can create a duplicate using another type of visualization.

3.7 Answers to Aviation Case Questions

1. What are the percentage delays by origin city?
   This visual shows that Orlando airport has 4436 delayed departure flights for 1772 days of the week. Tampa has the second-highest number of delayed flights (Fig. 3.14).
2. What are the arrival delays per flight date and the five numbers summary?
   If we select the flights for this date 01/07/2018, the box and whisker plot displays the min delays (26 min); the max delay is 134 min along with the average, the first quarter (Q1), and the third quarter (Q3). These can also be used to compare whether the distribution is symmetric or left/right-skewed (Fig. 3.15).

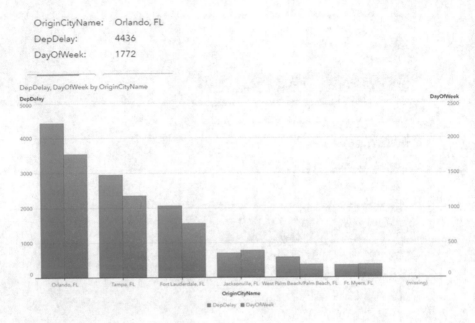

Fig. 3.14 Delays/origin and flight number

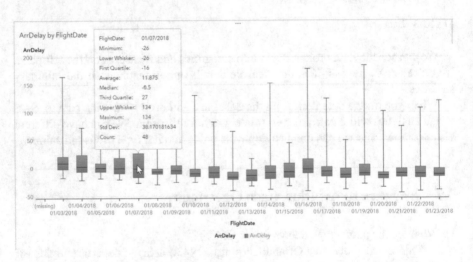

Fig. 3.15 Arrival delays/flight date

3. Which city has the highest frequencies of delays?

Within the Pie Chart, we can visually detect Orlando, followed by Tampa, with the highest frequencies of delays. Orlando has 462 delays, and 250 were related to NASdelay; 1441 were late aircraft delays, as presented in Fig. 3.16.

4. Which destination city had the highest traveled distance and delays?

In Fig. 3.17, we can visualize all flights with the highest distance and delays. Pointing to Islip, NYC, this city has the highest carrier delays (284 instances)

OriginCityName:	Orlando, FL
Frequency:	462
ArrDelay:	-315
DepDelay:	4436
WeatherDelay:	841
SecurityDelay:	0
NASDelay:	250
LateAircraftDelay:	1441
CarrierDelay:	1059

Frequency, ArrDelay, DepDelay, WeatherDelay, SecurityDelay, NASDelay, LateAircraftDelay, CarrierDelay by OriginCityName

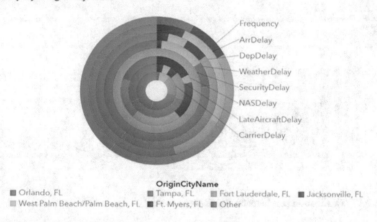

OriginCityName
■ Orlando, FL   ■ Tampa, FL   ■ Fort Lauderdale, FL   ■ Jacksonville, FL
■ West Palm Beach/Palm Beach, FL   ■ Ft. Myers, FL   ■ Other

Fig. 3.16 Frequencies of delays/city

with a traveled distance of 64,882. In contrast, Baltimore has the highest distance traveled but not the highest delays.
5. What is the total breakdown of the arrival delays with the week's days and the original city?

In this Fig. 3.18, we did create a summary table to summarize the arrival delays with the days of the week and the origin city. Orlando appears again as the city with the highest number of delays.
6. Can we detect any pattern within the frequencies of flights per day of the month, day of the week?

We did create a time series plot highlighting the delays per flight day of the month (second line in Fig. 3.19) and day of the week (third line in Fig. 3.19). The highest peak with arrival delays occurred on Jan 16.

Once you are done with your report, save it.

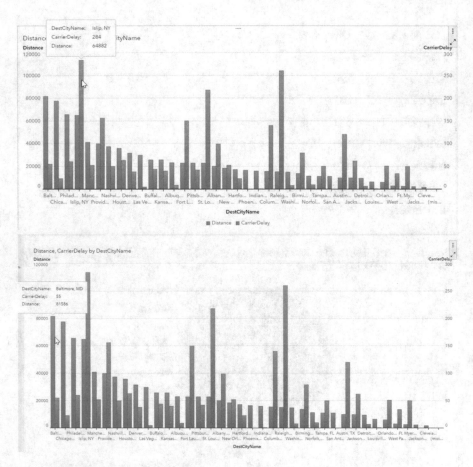

Fig. 3.17 Destination per distance and delays

ArrDelay ▲	DayOfWeek	OriginCityName
-315	1772	Orlando, FL
-154	388	Jacksonville, FL
65	183	Ft. Myers, FL
121	779	Fort Lauderdale, FL
253	184	West Palm Beach/Palm Beach, FL
528	1175	Tampa, FL

Fig. 3.18 Summary table arrival delays with days of the week and origin city name

Fig. 3.19 Time series plots

As we explained earlier in the chapter, knowing your data is the most critical step in the process. You can use graphical tools such as bar graphs, histograms, scatter plots, among other types, and numeric measures that include summary tables summarizing the measures of central tendency and measures of variation to explore your data. You can detect extreme values and outliers, identify the missing variables, and investigate data as counts or percentages form.

Understanding population behavior can make critical healthcare decisions more efficient and effective. Tectonics GEO hypothesized that Spring Breakers in South Florida contributed to the spread of COVID 19 along the east coast. The company isolated cell phone activity from just one beach in Fort Lauderdale in early March 2020 before social distancing became the order of the day. The company took the data from each device. It then analyzed secondary locations of the same mobile device to create a heat signature of where the mobile devices were located 2 weeks after the initial date. This heat map then provides public health officials with future hotspots of an outbreak. It allows them to pre-prepare health and testing facilities to receive sick individuals (Lewinski 2020).

References

Bureau   of   Transportation   Statistics   n.d.   https://www.transtats.bts.gov/databases. asp?Z1qr_VQ=E&Z1qr_Qr5p=N8vn6v10&f7owrp6_VQF=D

Bureau   of   Transportation   Statistics   (2018)   https://www.transtats.bts.gov/OT_Delay/OT_ DelayCause1.asp?20=E

Howard SK, Thompson K (2015-2016) Seeing the system: Dynamics and complexity of technolo-gyintegration in secondary schools. Education and Information Technologies 21(6):1877-1894. https://doi.org/10.1007/s10639-015-9424-2

Lewinski JS (2020, March 29) Controversial video claims to use spring break traffic to show dangers of not social distancing. Retrieved from https://www.forbes.com/sites/johnscottlewin-ski/2020/03/28/controversial-videoclaims-to-use-spring-break-traffic-to-show-dangers-of-not-social-distancing/#6313142b6ce7

Technews (2016) 5 Characteristics all excellent data visualization should have. https://ecmapping. com/2016/05/17/5-characteristics-all-excellent-data-visualization-should-have/

Yarmuluk D (2019) Discover the 4 key traits of a great data visualization. https:// www.microsoft.com/enus/microsoft-365/business-insights-ideas/resources/ discover-the-4-key-traits-of-a-great-data-visualization

Chapter 4
Evaluating Predictive Performance

Learning Objectives
- Learn the importance of evaluating predictive performance
- Selecting Algorithms
- Evaluate Predictive Performance

4.1 The Importance of Evaluating Predictive Performance

Analysts use predictive analytics to foresee if a change will help them reduce risks, improve operations, and/or increase revenue. At its heart, predictive analytics answers the question, "What is most likely to happen based on my current data, and what can I do to change that outcome?"

Predictive analytics uses data and statistical techniques, such as machine learning (ML) and predictive modeling, to forecast outcomes. The focus in predictive analytics is not on hypothesis testing but instead on detecting repeated patterns of values in the data that can be used to make accurate predictions of future outcomes.

Selecting the correct predictive modeling technique at the start of your project can save a lot of time. Choice of metrics influences how the predictive model's performance is measured and compared. But metrics can also be deceiving. If we do not use metrics that correctly measure how accurately the model predicts our problem, we might be fooled to think we built a robust model.

Now that you have ensured that you have enough appropriate data, massaged the data into a form suitable for modeling, identified key features to include in your model, and established how the model is to be used, you are ready to use powerful algorithms to build predictive models or discover patterns in your data. So what are algorithms?

L. Halawi et al., *Harnessing the Power of Analytics*, https://doi.org/10.1007/978-3-030-89712-3_4

4.2   Algorithms

Predictive analytics transform data into meaningful, usable, and actionable insights using algorithms.

Algorithms are step-by-step instructions to be executed to get the desired output or to model data. You can think of an algorithm like a cake recipe. If you want to make a cake, you follow a procedure to put the different ingredients together; see Fig. 4.1 highlighting the baking process and (Fig. 4.2) highlighting the steps for classifying data.

Data algorithms are used to determine models that fit the data or reliable patterns within them. They are relatively complex and draw on mathematical techniques from probability theory, information theory, estimation, uncertainty, cluster analysis, artificial intelligence, graph theory, and database techniques. The most popular algorithms are stated in Table 4.1.

Algorithms consist mainly of a specific mix of three components: the model, the preference criteria, and the search algorithm. There is no universally best method: choosing a particular algorithm for a specific application is something of an art, and consequently, skilled selections rarely run entirely automatically. This book, will cover only the models listed in Table 4.1.

There is no one answer or recipe in choosing the perfect algorithm. The advice is to try many different models and compare the results. The famous "no free lunch" theorem (Wolpert 1996) states that no one model works best for every problem.

You can, nevertheless, utilize the subsequent questions to inform your decision (Wujek et al. 2016):

- What is the size and nature of your data?

- If you anticipate a linear relationship between your features and your target, linear or logistic regression might be satisfactory and adequate.
- What are you trying to achieve with your model?
- State your question. For example, is it a classification to predict a value for an interval target, detect patterns or anomalies, or offer recommendations? Answering your question leads you to a suitable algorithm.

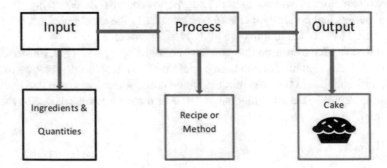

Fig. 4.1 Baking procedure

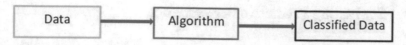

Fig. 4.2 Steps of classifying data

Table 4.1 Examples of algorithms and major uses

Algorithms	Major uses
Linear regression	1. To determine the strength of predictors 2. To forecast an effect 3. To trend forecasting
Logistic regression	1. To describe data 2. To explain the relationship between one dependent binary variable and one or more nominal, ordinal, ratio, or interval independent variables
Decision tree	1. To make a decision that uses probabilities to calculate likely outcomes
Random test	1. A random input is used to test the system's reliability 2. To test the system's performance
Neural network	1. Modeling nonlinear and nonlinearly separable phenomena 2. Fastest Scoring model
Forest and ensemble	1. Tree ensembles (forests, gradient boosting) can increase prediction accuracy and decrease overfitting

- How accurate does your model need to be?
- Simpler models train faster, are easy to understand, less prone to overfitting, and easy to deploy, making it easier to explain how, and why the results were achieved. Start with a simple model and walk your way toward more complex models
- How much time do you have to train your model?
- This is linked to how accurate your model needs to be. If you are pressed on time and need to train a model quickly, linear or logistic regression and decision trees are undoubtedly your best options.
- How interpretable or understandable does your model need to be?
- When interpretability or explainable documentation is critical, use decision trees or a regression technique
- Does your model have automatic hyperparameter tuning capability?

- If your model has hyper-tuning capabilities, leverage this capability.

Beginners are often inclined to choose algorithms that are easy to implement and can get results quickly. This tactic is adequate if it is the first step of the process. After you obtain some initial results and become acquainted with your data, you can experiment with more sophisticated algorithms. Most algorithms mainly involve careful tuning and extensive training to get the best possible performance. Picking the incorrect modeling technique may generate inaccurate predictions.

4.3   Evaluating Predictive Performance

We typically build several models, assess individual models, and compare them to determine the best model, commonly called a champion model. The champion model is then deployed into production, a process called scoring. After a model has been deployed, it is essential to monitor performance and update per requirements.

When we evaluate any model, we first consider the business needs and how critical they are, prediction capability, accuracy, speed to train, or the speed to score. The models are assessed in multiple stages: business evaluation, validation, and generalization to the population.

A predictive model can solve all problems that fall into two categories: a classification or regression problem. These categories will require the use of different metrics to evaluate your model. There are three types of outcomes of interest: predicted numerical value (price, height), predicted class membership/categorical value (sick, not sick), and propensity, that is, the probability of class membership when the outcome is categorical.

Prediction methods generate numerical predictions, while classification methods generate propensities and cutoff on these propensities. So how do we know whether our prediction is accurate and how we interpret the model and choose the best model?

4.3.1   Model Assessment

To choose the best model, you need to assess each model's performance. Evaluation metrics are essential for comparing across models, selecting the correct configuration of a specific model, and comparing to the baseline ("no model").

One or two of the most promising models are then selected from the contending models (using training data sets) and then applied to the validation and test samples. The validation results can be compared to help choose one or more final models. The developed models are assessed and evaluated for their accuracy and generality.

Validation methods such as Teach and Test Method, Running AI Model Simulations, and including overriding mechanism incorporate tests on the quality of the model. However, these methodologies are outside the scope of the book. The target may change dynamically in specific business scenarios, so we need to retrain the models quickly. In particular cases, the model may be scored in real time.

4.3.2   Metrics

There are several approaches for evaluating the model. Multiple methods are available to classify or predict. For each technique, numerous choices are available for settings. To choose the best model, one needs to assess each model's performance through the various metrics.

One must understand the performance of each of the models by picking metrics that truly measure how well each model achieves the overall business goals of the company.

For classification, metrics based on the confusion matrix include overall accuracy, specificity, sensitivity, and metrics that account for misclassification costs. For prediction, metrics include Average, Error, MAPE, and RMSE (based on the validation data).

4.3.2.1 Metrics for Classification Models

The following metrics can be used for evaluating a classification model:

Percent correction classification (PCC): measures overall accuracy without regard to what kind of errors are made; every error has the same weight.

Misclassification rate: this is one of the most widely used selection statistics, particularly when the target value is binary. It measures the proportion of misclassified data against the total data. Misclassification rate = (False positives + false negatives)/total instances). So basically, what we are looking for in this plot is how many observations were correctly classified and how many were classified incorrectly.

Confusion matrix: also measures accuracy but distinguishes errors, i.e., false positives, false negatives, and correct predictions. This matrix is a crosstabulation of the actual and predicted outcomes based on a decision rule. A confusion matrix displays four counts: true positives, true negatives, false positives, and false negatives. *Sensitivity*, the true positive rate, is the number of true positive decisions divided by the total number of known primary cases. *Specificity*, the true negative rate, is the number of true negative decisions divided by the total number of known secondary cases. These measures are the basis for the ROC chart, which you will learn about next. **Sensitivity (also called "recall")** = % of "C_1" class correctly classified. **Specificity** = % of "C_0" class correctly classified. **Precision** = % of predicted "C_1's" that are actually "C_1's".

These metrics are good to use when every data entry needs to be scored. If, however, you only need to act upon results connected to a subset of your data, you might want to use the following metrics:

Area Under the ROC Curve (AUC–ROC): is one of the most widely used metrics for evaluation. It is popular because it ranks the positive predictions higher than the negative. ROC is a graphic representation showing how a classification model performs at all classification thresholds. ROC Curve plots two parameters: True Positive Rate and False Positive Rate at different thresholds in the same graph. It tells how much the model is capable of distinguishing between classes. The ROC curve is plotted with TPR against the FPR, where TPR is on the y-axis, and FPR is on the x-axis. In general, an AUC of 0.5 suggests no discrimination, 0.7–0.8 is considered acceptable, 0.8–0.9 is deemed excellent, and more than 0.9 is deemed outstanding. The AUC value lies between 0.5 and 1, where 0.5 denotes a bad classifier, and 1 represents an excellent classifier. The Axis Y represents the cumulative rate of True Positives (sensitivity). Axis X represents the cumulative rate of False Positives

(1 – Specificity). The ROC curve represents the probability distribution of the detection and false alarms. The greater the area is above the random guess (diagonal), the better the model is.

Lift and gain charts: both charts measure the effectiveness of a model by calculating the ratio between the results obtained with and without the predictive model. In other words, these metrics examine if using predictive models has any positive effects or not. Cumulative gains and lift charts are visual aids for measuring model performance. Both charts consist of a lift curve and a baseline.

The lift chart measures how much better one can expect to do with the predictive model compared without a model. Lift is calculated as the ratio between the results obtained with and without the predictive model. The higher the area between the lift curve and the baseline, the better the model. Lift helps you decide which models are better to use.

The cumulative gains chart shows the percentage of the overall number of cases in a given category "gained" by targeting a percentage of the total number of cases. Gain at a given decile level is the ratio of the cumulative number of targets (events) up to that decile to the total number of targets (events) in the entire dataset. The diagonal line is the "baseline" curve; the farther above the baseline a curve lies, the greater the gain.

4.3.2.2   Metrics for Regression Models: Numerical Values

Numeric measures used to evaluate model performance are named assessment measures or fit statistics. To know which assessment measure one should use, you select a proper assessment measure based on the following two factors: the target measurement scale or the prediction type. We want to know how well the model predicts new data, not how well it fits the data it was trained with.

A regression problem is about predicting a quantity. To evaluate how good your regression model is, you can use the following metrics:

- **R-squared**: is a measure of how close the data are to the fitted regression line. It is also known as the coefficient of determination or the coefficient of multiple determination for multiple regression. R-squared does not take into consideration any biases that might be present in the data. An R^2 of 1 indicates that the regression predictions perfectly fit the data. R-Squared does not penalize for adding features that add no value to the model. So an improved version over the R-Squared is the **adjusted R-Squared.** Adjusted R-squared more than 0.75 is a very good value for showing the accuracy. In some cases, an adjusted R-squared of 0.4 or more is acceptable as well.
- **Mean square error (MSE)**: good to use if you have many outliers in the data.
- **Mean absolute error MAE or MAD**: Mean absolute error (deviation): similar to the average error, only you use the absolute value of the difference to balance out the outliers in the data, Gives an idea of the magnitude of the average absolute errors.

- **Mean error is similar to MAE** but retains the errors, indicating whether the predictions are averaged over or underpredicting the outcome variable.
- Mean absolute percentage error (**MAPE**): It measures this accuracy as a percentage and can be calculated as the average absolute percent error for each period minus actual values divided by actual values. The MAPE is also sometimes reported as a percentage. It is irresponsible to set arbitrary forecasting performance targets (such as **MAPE** < 10% is Excellent, **MAPE** < 20% is **Good**) without the context of the forecastability of your data.
- **RMSE** (root-mean-squared-error): Square the errors, find their average, take the square root. For a datum that ranges from 0 to 1000, an **RMSE** of 0.7 is small, but if the range goes from 0 to 1, it is not that small anymore. When the RMSE decreases, the model's performance will improve. Based on a rule of thumb, it can be said that RMSE values between 0.2 and 0.5 show that the model can relatively predict the data accurately.
- **Total SSE**: Total sum of squared errors. SSE is the sum of the squared differences between each observation and its group's mean. It can be used as a measure of variation within a cluster. If all cases within a cluster are identical, the SSE would then be equal to 0.

4.3.3   Selecting Model Fit Statistics by Prediction Type

Model fit statistics can be grouped by prediction type as presented in Fig. 4.3.

For *decision* predictions, the model comparison tool rates model performance based on accuracy or misclassification, profit or loss, and by the Kolmogorov–Smirnov (KS) statistic.

Accuracy and misclassification tally the correct or incorrect prediction decisions. The Kolmogorov–Smirnov statistic describes the ability of the model to separate the primary and secondary outcomes. The Kolmogorov–Smirnov (Youden) statistic is a goodness-of-fit statistic that represents the maximum distance between the model ROC curve and the baseline ROC curve. The maximum value of the **Youden index** is 1 (perfect test), and the minimum is 0 when the test has no diagnostic value.

For *ranking* predictions, two closely related measures of model fit are commonly used. The ROC index is like concordance (described above in Sect. 4.3.2.1). The ROC index equals the percent of concordant cases plus one-half times the percent of tied cases. The Gini coefficient (for binary prediction) equals 2* (ROC Index − 0.5). **Gini** above 60% is a **good model**. *The **Gini index** varies between values 0 and 1, where 0 expresses the purity of classification, that is, All the elements belong to a specified class or only one class exists there. And 1 indicates the random distribution of elements across various classes. The value of 0.5 of the Gini Index shows an equal distribution of elements over some classes.*

For *estimated* predictions, there are at least two commonly used performance statistics. Schwarz's Bayesian criterion (SBC) is a penalized likelihood statistic. This likelihood statistic can be thought of as a weighted average squared error.

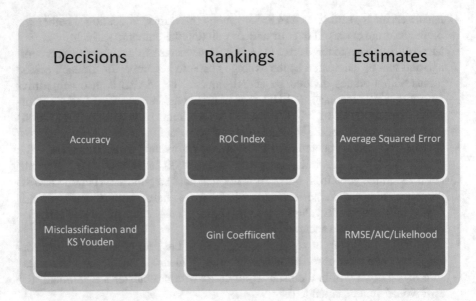

Fig. 4.3   Model selection statistics

4.3.3.1   Fit Statistics

For assessing prediction performance, several measures are used. In all cases, the estimates should be based on the validation set. Based on the training set, errors in general (review Sect. 4.3.2.2) tell us about the model fit, while those based on the validation data set measure the model's ability to predict new data. We expect training errors to be smaller than the validation errors—the more complex the model, the higher chance it will overfit the training data.

The test data set is used to evaluate how the model would perform based on new data. It is useful to assess how the model can generalize to new data, which most likely will differ from the data used to train and validate the model.

The two-sample Kolmogorov–Smirnov (KS) test is a nonparametric test that compares the cumulative distributions of two datasets (1,2) and reports the maximum difference between the two cumulative distributions and calculates a P-value from that and the sample sizes. **K-S** should be a high value (Max = 1.0) when the fit is good and a low value (Min = 0.0) when the fit is not good. When the K-S value goes below 0.05, you will be informed that the Lack of fit is significant.

The Gini Coefficient or Gini Index measures the inequality among the values of a variable. Higher the value of an index, the more dispersed the data is. Alternatively, the Gini coefficient can also be calculated as half of the relative mean absolute difference. The Gini coefficient is usually defined mathematically based on the Lorenz curve, which plots the proportion of the population's total income (y-axis) cumulatively earned by the bottom $x\%$ of the population. Another way of thinking about the

Gini coefficient is as a measure of deviation from perfect equality. The way the index is constructed implies that lower values are an indication of higher levels of equality. Other things being the same, this is a good thing.

References

Wolpert D (1996) The lack of a priori distinctions between learning algorithms. Neural Comput 8(7):1341–1390

Wujek B, Hall P, Güneş F (2016) Best practices for machine learning applications. SAS Institute Inc., Cary

Websites

https://towardsdatascience.com/classifying-rare-events-using-five-machine-learning-techniques-fab464573233

https://towardsdatascience.com/clearly-explained-gini-coefficient-and-lorenz-curve-fe6f5dcdc07

https://www.sciencedirect.com/topics/engineering/misclassification-rate

https://machinelearningmastery.com/data-preparation-is-important/

https://www.netsuite.com/portal/resource/articles/financial-management/predictive-modeling.shtml

https://indatalabs.com/blog/predictive-models-performance-evaluation-important?cli_action=1621699331.16

https://towardsdatascience.com/selecting-the-correct-predictive-modeling-technique-ba459c370d59

https://medium.com/analytics-steps/understanding-the-gini-index-and-information-gain-in-decision-trees-ab4720518ba8

https://www.analyticsvidhya.com/blog/2019/08/11-important-model-evaluation-error-metrics/

https://www.listendata.com/2014/08/excel-template-gain-and-lift-charts.html

http://www2.cs.uregina.ca/~dbd/cs831/notes/lift_chart/lift_chart.html

https://www.ibm.com/docs/es/spss-statistics/24.0.0?topic=overtraining-cumulative-gains-lift-charts

Chapter 5
Decision Trees and Ensemble

Learning Objectives
- Overview of decision trees
- Identify different terms used with decision trees
- Explain the math behind decision trees
- Explain overfitting and pruning
- Application example: decision tree
- Explain ensemble

5.1 Overview of Decision Trees

Decision trees are among the most popular, flexible, and reliable data mining methods for developing predictive models. They are flexible in that they can model interval (regression trees), ordinal, nominal, and binary (classification trees) targets and accommodate nonlinearity and interactions. They are the easiest to build, understand, and interpret. Decision trees are not affected by missing variables, we do not have to impute data, and they are not affected by outliers. Also, we do not need to normalize or scale features, and they can be built interactively, or system generated.

Decision Trees are represented as a tree-like structure asking questions to partition (split) the data. The model is read from top-down, starting at the root node. The critical step in describing the tree is to settle on the appropriate question for the root and consequently for all the branches. At the root of the tree, partitioning is more straightforward. The objective is to find the model that will generalize well, not too basic or too complicated.

To develop a tree model, we need two datasets. The training data set is used for creating the initial model, and the validation data set is used for pruning or

© The Author(s), under exclusive license to Springer Nature
Switzerland AG 2022
L. Halawi et al., *Harnessing the Power of Analytics*,
https://doi.org/10.1007/978-3-030-89712-3_5

fine-tuning the model. The test dataset is an optional additional assessment used when you need to compare your model's performance to the performance of other models.

Decision tree algorithms are called CART (Classification and Regression Trees) created by Breiman et al. (1984). We can build a decision tree model to predict a categorical variable; we call it a classification tree. On the other hand, with a continuous variable, we create what we call a regression tree model. In supervised learning, the target result is already known.

5.1.1   Classification Trees

A classification tree is a graphical representation of rules for classifying observations into two or more groups. These trees use a hierarchical classifying process that entails splitting nodes and terminal nodes that categorize datasets into homogenous groups. As the term implies, the decision tree is where each branch denotes a particular classification question, and all leaves are partitions. The goal is to build a model that predicts the value of a target variable based on several input variables.

A decision tree or a classification tree is a tree in which each internal (non-leaf) node is labeled with an input feature. It is a (Yes/No) type where the outcome is a variable like "happy" or "unhappy." The decision variable is Categorical.

The tree is built through a process known as binary recursive partitioning and is the standard method to fit decision trees. Recursive partitioning is an iterative process in which data is split into smaller and smaller subsets. Classification tree algorithms generate choices for the independent variable and which splitting value should be picked by minimizing the average weighted impurity of the resulting partitions. The output of the classification problem is taken as "Mode" of all observed values of the terminal node. The application of classification includes fraud detection, medical diagnosis, and target marketing.

5.1.2   Regression Trees

The decision tree is called a regression tree with a continuous variable, such as Height <180 and Loan >30,000. The leaves give the predicted value of the target. The path to every leaf is based on a Boolean Rule.

We reduce the variance with a decision tree by finding subsets of data with reduced variation and then taking the average target variable within the group. We can reuse the predictors within each split. The output of regression analysis is the "Mean" of all observed node values.

However, it is essential to note that increasing the number of divisions does not guarantee to improve the model and may make it harder to train.

5.2   Terms Used with Decision Trees

We need to know the terminology used to work with decision trees, whether as a classification example or regression model.

1. Root node represents the entire population or sample. We start dividing the data into two or more homogeneous sets. The input variable is the explanatory variable.
2. Splitting is the process of dividing a node into two or more subnodes. Splits that enhance the purity of a node and reduce the randomness are informative.
3. Decision node is the process when a subnode splits into additional subnodes.
4. Leaf/terminal node does not partition further. Terminal nodes are called leaves and represent the predicted target.
5. Pruning is the process that follows the training phase. First, we allow the tree to grow without applying any explicit restrictions. Then, we remove subnodes that are not populated enough.
6. Parent and child node is a node divided into subnodes called the parent node of subnodes.

5.3   The Math Behind Decision Trees

When building a decision tree, we may have many considerations, such as how many questions we have, how many splits we will do, and when we stop. The choice of input variables is based on statistical tests. The variable with the most significant influence is selected first. This section aims to illustrate some of the math used in decision trees and describe what happens in the backend of the decision trees algorithms.

5.3.1   Expected Value

An expected value (EV, expected average) is a long-run average value of random variables, indicating the probability-weighted average of all possible values.

If one event is repeated n times, then EV = $P(x) \times n$. Where EV is the expected value, $P(x)$ is the probability of the event x, and n is the number of repetitions of the event x. There are many problems related to the expected value involving multiple events, for instance, finance problems. In this case, EV is the probability-weighted average for all possible activities. Hence, the general formula for various events is:

$$EV = \sum_{i=1}^{n} p(x_i) \times x_i$$

5.3.2   Measure of Level of Impurity

The purity of a set of examples is the homogeneity of its cases regarding their classes. A decision tree splits on nodes made according to impurity metric.

In Fig. 5.1 below, there is an example of the different levels of impurity between the buttons and soccer balls.

The most common ways to measure impurity are the GINI Index and the Information Gain. We describe both below.

5.3.3   Attribute Selection Measures

Attribute selection measures help decide which feature leads to the purest subset. It determines how the data at a given node are to be split. To have the best spilt, the selected quality measure should be calculated for all features.

This chapter introduces three popular attribute statistics measures (ASM) used to make precise decisions: Information Gain, GINI Index, and Gain Ratio. When training a Decision Tree using these metrics, the best split is chosen by maximizing IG.

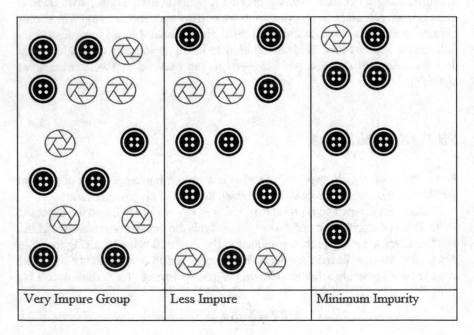

Fig. 5.1 Levels of impurity

5.3.3.1 Information Gain

The information gain (IG) is information gained about a random variable or signal from observing another random variable. It measures how important a given attribute is. The measure is used to decide the order of this attribute in the node of a decision tree. (IG) is key to constructing a decision tree. Therefore, the decision tree algorithm always works on maximizing IG. As a result, an attribute with the highest (IG) will be tested. IG is calculated for a split by subtracting the weighted entropies of each branch from the original entropy.

Entropy is the average rate at which a stochastic data source produces information. Entropy is the measure of impurity, uncertainty in a bunch of examples. It controls how a decision tree decides to split the data. The Entropy measure ranges between 0 (when all the observations belong to the same level) and 1 (when all n groups are represented in equal proportions). Suppose that (P_i) is the probability of a class (i), and n is the number of levels; the IG gain can be calculated as follows (Fig. 5.2):

Example

In this example, we have two classes, the black icon and the green one. The probability of the black icon is 8/14, and the green icon is 6/14. Therefore, we can compute the Entropy measure as follows (Fig. 5.3).

$$\text{Entropy} = -8/14\left(\log_2 8/14\right) - 6/14\left(\log_2 6/14\right)$$
$$= -8/14(-0.7) - 6/14(-1.32)$$
$$= -0.6(-0.7) - 0.4(-1.32)$$
$$= 0.42 + 0.528$$
$$\approx 0.948. \left(\text{The higher the entropy, the more the information Content}\right).$$

The entropy of minimum impurity is 0. An example in Fig. 5.4 is below:

Hence, Entropy $= -1(\text{Log}_2 1) = 0$ where $\text{Log}_2 1 = 0$. It is important to note that this is not a good learning example.

The entropy of a group consists of two classes, where each one represents 50% in the group, as presented in Fig. 5.5.

The probability (P$_i$) is the probability of a class (i), $P_i = \dfrac{\#Samples\ in\ subject(children)}{\#Samples\ in\ dataset\ (parent)}$,
The entropy formula is $I = -\sum_{i=1} p_i \log_2(p_i)$,
Information Gain formula is **IG= Entropy(parent)- [Average Entropy(children)]**.

Fig. 5.2 How to calculate the information gain

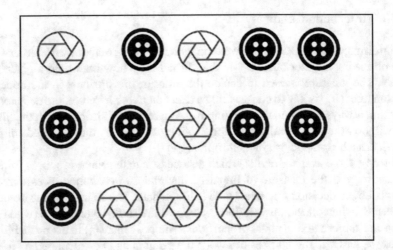

Fig. 5.3 Impure group

Fig. 5.4 Entropy for min
impurity

Fig. 5.5 Entropy group

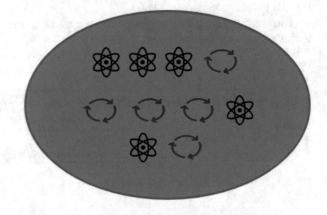

$$\text{Entropy} = -1/2\left(\log_2 1/2\right) - 1/2\left(\log_2 1/2\right)$$
$$= -0.5(-1) - 0.5(-1) \quad \text{where} \quad \log_2 0.5 = -1)$$
$$= 1\left(\text{This is an excellent training set for learning}\right).$$

Hence, the higher the entropy, the more impure a dataset, and the less impure the dataset, the lower the entropy.

How to Calculate Information Gain

Let us assume we have a population of 22 instances with two classes, respectively, represented as (◆) and (➤) in Fig. 5.6.

Let us first calculate the probabilities. We have $p($ ◆ $) = 14/22$ and $p($ ➤ $) = 8/22$. Hence,

$$\text{Entropy}\left(\text{parent}\right) = -14/22\left(\log_2 14/22\right) - 8/22\left(\log_2 8/22\right)$$
$$= -0.64\left(\log_2 0.64\right) - 0.36\left(\log_2 0.36\right)$$
$$= -0.64(-0.64) - 0.36(-1.47)$$
$$\approx 0.94$$

$$\text{Now, the Entropy}\left(\text{child1}\right) = -10/12\left(\log_2 10/12\right) - 2/12\left(\log_2 2/12\right)$$
$$= -0.83\left(\log_2 0.83\right) - 0.17\left(\log_2 0.17\right)$$
$$= -0.83(-0.27) - 0.17(-2.56)$$
$$= 0.66.$$

$$\text{The Entropy}\left(\text{child2}\right) = -4/10\left(\log_2 4/10\right) - 6/10\left(\log_2 6/10\right)$$
$$\approx -0.4\left(\log_2 0.4\right) - 0.6\left(\log_2 0.6\right)$$
$$\approx -0.4(-1.32) - 0.6(-0.74)$$
$$= 0.97.$$

$$\left(\text{Weighted}\right)\text{Average entropy}\left(\text{children}\right)$$
$$= P\left(\text{child1}\right).\text{Entropy}\left(\text{child1}\right) + P\left(\text{child2}\right).\text{Entropy}\left(\text{child2}\right)$$
$$= 12/22(0.66) + 10/22(0.97)$$
$$\approx 0.8.$$

$$\text{Then,}\left(IG\right) = \text{Entropy}\left(\text{parent}\right) - \left[\text{Average Entropy}\left(\text{children}\right)\right].$$
$$= 0.95 - 0.8 = 0.15. \text{Hence, } 0.15 \text{ gains information for this split.}$$

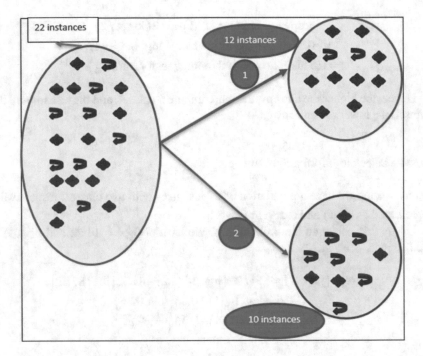

Fig. 5.6 Calculating information gain

5.3.3.2   Gini Index

Gini Index measures how often a randomly chosen element would be incorrectly classified. It works with categorical target variables "Success" or "Failure," which performs only binary splits. The degree of the Gini Index varies between 0 and 1. Hence, we should choose the attribute with the least Gini Index as the root node to build a better decision tree.

The Gini Index measures the impurity of data. Let (P_i) be the probability of an object being classified to a particular class and n the number of objects in this class. If all the partition (i) observations belong to the same group, its Gini index value will be 0, indicating perfect purity. Alternatively, when all n groups are represented in equal proportion in the partition (i), its GINI index will reach its maximum value $(n - 1)/n$. Figure 5.7 presents the formula to calculate the GINI index.

GINI Index Example

The data for this example is associated with direct marketing campaigns (phone calls) of a Portuguese banking institution. The classification goal is to predict if the client will subscribe (yes/no) a term deposit (variable y) to a term deposit (variable y) (Moro et al. 2014).

The GINI index for a given partition (i) is defined as Gini(D)= $1 - \sum_{i=1}^{n} p_i^2$.

If a data set D is split on A into two subsets D_1 and D_2, the Gini Index, Gini (D) is defined as $Gini_A(D) = \frac{|D_1|}{|D|}$ Gini $(D_1) + \frac{|D_2|}{|D|}$ Gini (D_2).

Fig. 5.7 How to calculate the GINI index?

We take a sample of 500 people from the dataset Bank (Moro et al. 2014) with two variables age (less or equal than 30/more significant than 30) and Marital Status (married/single and divorced). Then, we segregate the people based on the target variable (having a loan or not). We have 83 out of 500 who have loans, and 54 are married. Then, we want to identify which split is producing more homogeneous subnodes by using Gini. Figure 5.8 represents the split in Age.

Calculate, Gini for subnode people aged $\leq 30 = 1 - [(0.18^2 + 0.82^2)] \approx 0.3$ where P(people aged ≤ 30 and having loans) $= 12/68 \approx 0.18$ 2.

Calculate, Gini for subnode People aged $>30 = 1 - [(0.16^2 + 0.84^2)] \approx 0.27$ where p(people aged >30 and having loans) $= 71/432 \approx 0.16$

Calculate Weighted Gini for split on age $= 68/500(0.3) + 432/500(0.27) \approx 0.27$. Now, Fig. 5.9 represents a split on marital status.

1. Calculate, Gini for subnode Married $= 1 - [(0.18^2 + 0.82^2)] \approx 0.30$ where p(people married and having loans) $= 54/307 \approx 0.18$.
2. Calculate, Gini for subnode single and divorced $= 1 - [(0.15^2 + 0.85^2)] \approx 0.26$ where p(people not married and having loans) $= 29/193 \approx 0.15$.
3. Calculate   weighted   Gini   for   split   on   marital $= 307/500(0.30) + 193/500(0.26) = 0.28$.

In conclusion, we can see that the Gini score for splitting among Ages is higher than splitting on Marital Status. Hence, the node split will take place on Age. SAS confirms the above result; we can see the variable importance in this order month, Age, $Q(y)$, and marital status in Fig. 5.10.

5.3.3.3   Gain Ratio

Gain Ratio modifies the information gain by reducing its bias. It considers the number and size of branches when choosing attributes and introduces the entropy concept. The attribute with maximum Gain Ratio is selected as the splitting attribute. Figure 5.11 highlights the calculations for the Gain Ratio.

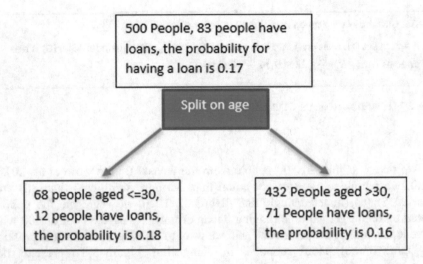

Fig. 5.8 Split on age

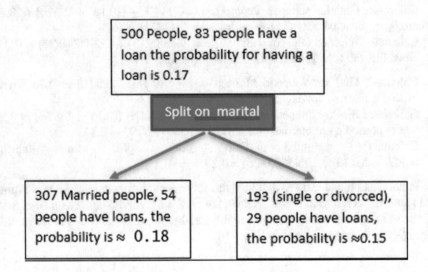

Fig. 5.9 Split on marital status

5.4   Avoiding Overfitting

Overfitting refers to a model that models the training data too well. Overfitting occurs when a model learns the detail and noise in the training data to the extent that it negatively impacts performance when scoring new data. Some of the causes of overfitting are noise, lack of samples, and multiple comparison procedures.

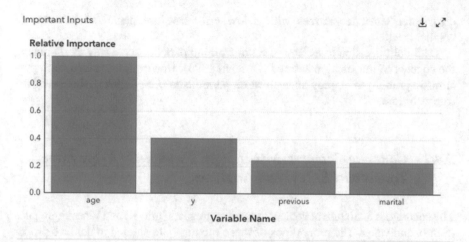

Fig. 5.10 Plot—relative importance for input variables

$$\text{Gain Ratio} = \frac{Information\ Gain}{Entropy} = \frac{Entopy(parent) - [Average\ Entropy\ (children)}{-\sum_{i=1}^{n} p_i log_2(p_i)}$$

Fig. 5.11 How to calculate the gain ratio?

One way to avoid overfitting is to use an ensemble. Ensemble methods combine numerous decision trees to generate better predictive performance than a single decision tree. The ensemble model's key principle is that a collection of weak learners come together to create a strong learner. Occasionally, the number of nodes is reduced considerably, leading to an underfit tree. Moreover, they can result in unstable models as minor changes with the training data may result in more significant changes in the tree's topology. Another popular way to avoid overfitting is pruning.

5.4.1   Pruning

Pruning involves removing the branches that make use of features having low importance. This way, we reduce the tree's complexity, thus increasing its predictive power. So with pruning, we limit the full-growing of a tree to a level where it does not overfit. Pruning can start at either root or the leaves.

Two commonly applied approaches for finding the best subtree are cost-complexity pruning (Breiman et al. 1984) and C4.5 pruning (Quinlan 1993).

Cost-complexity pruning is a widely used pruning method initially proposed by Breiman et al. (1984). The cost-complexity pruning method helps to prevent overfitting by making a trade-off between the complexity (size) of a tree and the error rate; Thus, large trees with a low error rate are penalized in favor of smaller trees.

C4.5 built decision trees from a set of training data the same way as ID4 using the concept of information entropy (Quinlan 1993). However, it evolved from pessimistic pruning to employ an even more pessimistic (i.e., higher) estimate of the true error rate.

5.4.2   Concrete Compressive Strength Example 1: Regression Tree Model SAS Visual Analytics

The concrete is a mixture of cement, coarse aggregates (gravel) and Fine Aggregates (Sand), and water. The proportions of these mixtures are studied to achieve proper compressive strength. Engineers use a superplasticizer while mixing the components to improve the flowability of concrete. The strength of the concrete is achieved by studying the proportions of the mixes.

In some cases, fly ash can add to the concrete's final strength and increase its chemical resistance and durability. Fly ash can significantly improve the workability of concrete. Granulated blast furnace slag has been used as a raw material for cement production. As an aggregate and insulating material, granulated slag has also been used as sandblasting shot materials.

How do you decide about the right or the specific composition of concrete? This is where the concrete compressive strength (f prime C) is of particular significance to the engineer.

How much strength the concrete has? If you add more cement, the mix would be stronger up to a certain proportion. As cement is the main factor and makes the particles hold, the Sand can help to fill the void in the concrete mix as you should not leave a void in the concrete; gravel takes the load and water (makes the paste that glues everything together). Superplasticizer is an addition to make the concrete flowable. It does not weaken cement, unlike adding more water that might weaken the concrete.

The dataset used in this example comprises 1030 instances with nine attributes consisting of eight quantitative input variables and one quantitative output variable (Refer. We do not have any missing values). Concrete is an important material in civil engineering. The concrete compressive strength is a highly nonlinear function of age and ingredients, and we will use it as our target variable (Yeh 1998) (refer to Appendix D).

For this demonstration, we are looking at answering the following question: What determines the compressive strength of concrete? Additional questions that one may consider if data is available are: What are the other critical parameters for concrete strengths other than compressive strength?

The dataset with the properties is as follows (Table 5.1).

Table 5.1 Data set properties table

	Variable Name ↑	Label	Type	Role
☐	Age (day)		Numeric	Input
☐	Blast Furnace Slag (component 2)		Numeric	Input
☐	Cement (component 1)		Numeric	Input
☐	Coarse Aggregate (component 6)		Numeric	Input
☐	Concrete compressive strength		Numeric	Target
☐	Fine Aggregate (component 7)		Numeric	Input
☐	Fly Ash (component 3)		Numeric	Input
☐	Superplasticizer (component 5)		Numeric	Input
☐	Water (component 4)		Numeric	Input

Using SAS VIYA Visual, we create a decision tree with a target variable: concrete compressive strength, and then add the attributes we want to add to our model.

Notice that SAS visual statistics quickly grows and prunes a decision tree. It also identifies which attributes are important in general and which attributes are not crucial to compressive Strength (Fig. 5.12). The modeling environment automatically splits the data according to our partitioning configuration (60 for training and 40 for validation), applying the auto-tuning feature dealing with missing values, and creating a decision tree structure along with all the relevant model fit statistics. It is an excellent common practice to start with this option and work yourself down to see if you can beat this option. Let us now look at what was created and what insights we can extract from our decision tree model.

There are several charts and plots to help you evaluate the model's performance.

The first plot is the *Tree Diagram*, which presents the final tree structure for this model, with the depth of the tree and all end leaves (Fig. 5.13).

The decision tree is drawn upside down with its root (Age) at the top, it is a series of decision rules, and the values within the decision rules may appear several times. In the image on the left, the bold text in black represents a condition/**internal node (cement component 1, water, etc.),** based on which the tree splits into branches/**edges**. The end of the branch that doesn't split anymore is the decision/**leaf**. If a variable is used in an earlier split (such as age, cement component 1, water) near the top of the tree, it is more important because it contributes more to the target event. This aligns well with the analysis in the variable importance plot.

So Age, cement component 1, and water contribute the most in predicting the compressive strength of cement. The color of the nodes in the plot indicates the predicted level of that node (teal for "high" and orange for "low").

Finally, the Output window shows the final decision tree model parameters, the Variable Importance table, and the pruning iterations. The variable importance plot ranks the attributes based on the contributions to the splits in the entire tree. If an

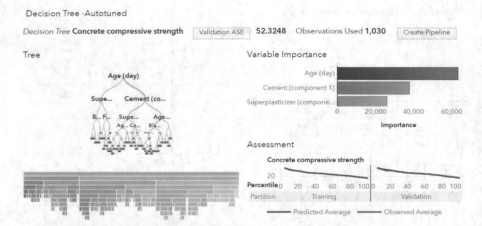

Fig. 5.12 DT concrete compressive strength

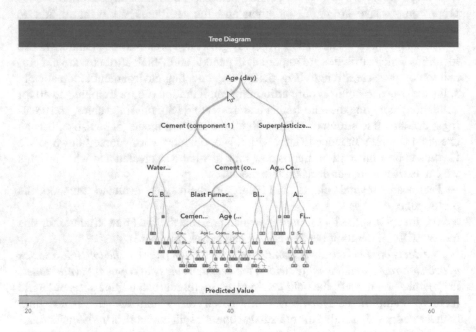

Fig. 5.13 Compressive strength autotuned decision tree diagram

attribute is used in an earlier split, it contributes more to the target variable. In this example, cement component 1, age, and water are the most important variables.

Six hundred eighteen observations were used with the training dataset. The maximum depth of the tree was 10, we had 90 leaves before pruning, and 89

The SAS System

The TREESPLIT Procedure

Model Information	
Split Criterion	VARIANCE
Pruning Method	Cost Complexity
Max Branches per Node	2
Max Tree Depth	10
Tree Depth Before Pruning	10
Tree Depth After Pruning	10
Number of Leaves Before Pruning	90
Number of Leaves After Pruning	89

	Training	Validation	Total
Number of Observations Read	618	412	1030
Number of Observations Used	618	412	1030

Fig. 5.14 Treesplit procedure

				Fit Statistics							
Target Name	Data Role	Partition In...	Formatted ...	Sum of Fre...	Average Sq...	Divisor for ...	Root Avera...	Mean Abso...	Root Mean ...	Mean Squa...	Root Mean ...
Concrete compressive strength	VALIDATE	0	0	412	43.1085	412	6.5657	5.0706	2.2518	0.0479	0.2189
Concrete compressive strength	TRAIN	1	1	618	24.2597	618	4.9254	3.6933	1.9218	0.0223	0.1493

Fig. 5.15 Fit statistics

leaves after pruning, and Cost Complexity was the pruning method used (Fig. 5.14).

Our model has a validation MSE of 0.0479, meaning that, on average, our model has produced predictions that are off from the actual value by 0.0479. One of our model tuning steps would be to reduce this value to predict the compressive strength with higher accuracy and produce a better fit model (Fig. 5.15).

5.5   Ensemble Methods Explained

An ensemble method is the grouping of predictive functions derived from different models. An ensemble is centered on the powerful idea of combining models. We have two types of ensemble methods: parallel and sequential. We start training different models on our data with the parallel type, combining them somehow. With sequential, you fit a model, you look at your results and then see how it does and adjust things, and then you fit another model on a modified data set; in some way, it is not the same model exactly and maybe do another one. Then in each step, you are fixing the errors from previous steps. So, in the end, you add together with the predictive functions where each step was trained to reduce the error of the previous steps.

Ensembles are used for prediction modeling; the prediction can be either classification or regression/estimation type, clustering, and association rule mining. They can be used for supervised and unsupervised machine learning tasks. Ensembles produce better accuracy and more stable/robust and reliable outcomes. Ensemble modeling can provide a form of relief as your predictions result from a collective effort or consensus amongst several trained models. Ensembles are not suitable for every solution; their pertinence is established by the modeling objectives described during the problem definition.

Ensemble models have three basic components: the dataset where it is jumbled and sent to the machine learning model; the mixing of the data is known as sampling, and we can perform either a row or column sampling. Next, we have a group of base learners, i.e., machine learning models that can be either dependent or independent, where each base model gives a prediction based on the data fed to it. Lastly, the output is the prediction of all the base models is combined in a final model. Here the decision is made based on the majority vote or by looking at the predictions weights.

Model Ensembles can be classified into two dimensions. The ensembles' method type may be categorized into bagging or boosting, and the model type, where the ensembles may be classified into homogeneous or heterogeneous types. Homogenous type combines the outcomes of two or more of the same models, such as decision trees and random forest, and uses Bagging or boosting. Heterogeneous model ensembles combine the outcomes of two or more different types of models such as decision trees, artificial neural networks, logistic regression, and others.

One of the most important terms with ensemble models is bias and variance. Bias is useful to quantify how much, on average are the predicted value is different from the actual value; the high bias error means that we have an underperforming model which keeps missing essential trends. On the other hand, variance quantifies how much of the predictions made on the same observation are different. A high variance model will overfit your training data and perform poorly on any observation beyond the training data set.

5.6   Ensemble Methods

There are several ensemble learning models: Bagging, Boosting, and Stacking (voting).

5.6.1   Bagging or Bootstrap Aggregation

Bagging, a parallel ensemble method, is also known as Bootstrap AGGregation. It is the simplest and most common ensemble method. Boostrap establishes the foundation of bagging techniques. It is a sampling technique where we select an observation out of the population of n observations, but it is entirely random. So, each observation can be chosen from the original population so that it is equally likely to be selected in each iteration of the bootstrapping process. After the bootstrap samples are formed, separate models are trained with the bootstrap samples where these samples are drawn from the training datasets, and sub-models are tested using validation or testing datasets. The final output prediction is combined across a projection of all sub-models.

The idea behind it is simple yet powerful; you build multiple decision trees from resampled data and noisy data and combine the predicted values through averaging or voting by creating a model with low variance. Bagging increases performance stability and prevents overfitting by distinctly modeling diverse data samples and then combining them. It is favorable for algorithms such as trees and neural networks. While the Bagging was initially established for decision trees, it can be used with any predictive model algorithm that yields outcomes with enough variation in the predicted values for both classification and regression/estimation type prediction problems.

5.6.1.1   Bootstrap Technique

Bootstrap is an ML ensemble mega-algorithm for reducing bias and variance. Bias is introduced when the machine learning method cannot capture the real relationship. The difference in fits between datasets is called variance. Highly flexible algorithms usually have low biases but high variance.

The following presents an explanation of this process. Let D be the training data with n training instances. Then, take repeated Bootstrap samples from this training set D with a replacement where the bootstrap sample D' with m values is less than n. For each sample D'_i corresponds a model M_i. Therefore, we have k models $M_1, M_2, ..., M_k$.

Now, suppose that y_i is the output from the module M_i (Fig. 5.16). We can then fit a weak learner for each of these samples and lastly group them by majority vote or average.

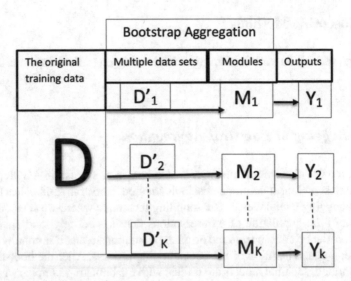

Fig. 5.16 Bootstrap aggregation

We use the weight for the outputs when Bagging is used for regression and the majority for the classification/qualitative response. For getting an ensemble model with low variance, for instance, we can define our strong model such that

The simple average, for regression problem, denoted by $S_L(.) = \frac{1}{L}\sum_{l=1}^{L} W_l(.)$.

The simple majority vote, for classification problem, denoted by $S_L(.) = \arg_k \max[\text{card}(l/W_l(.) = k]$.

Where L is the number of weak learners.

A model has high variance if the model changes a lot with changes in training data. Bagging is a concept to reduce variance in the model without impacting bias.

5.6.1.2   Random Forest

Random forest is generally the most prevalent bagging technique. It can also be used for classification and regression as well. When we use different learning models, we increase the classification (Bagging) accuracy. Random forest algorithm works as an extensive collection of de-correlated decision trees to make a better classification.

The Random Forest algorithm starts with selecting random samples from a given dataset. This algorithm will then construct a decision tree for every instance, get the prediction result from every decision tree, and then votes for every predicted result. At last, it selects the most voted prediction result as the final prediction result.

5.6.2 *Boosting*

The second most common ensemble method uses a set of algorithms that combine weak learners to form strong learners and thus increase the model's accuracy. Here, we first build a simple classification model slightly better than the random chance. It is a relatively simple variation of bagging where the different combined weak models are no longer fitted independently.

Here weak learners are sequentially produced during the training phase. The model's performance is improved by assigning a higher weightage to the previous incorrectly classified samples.

In boosting, we fit a model to the data, draw a sample, so that misclassified records have higher selection probabilities, fit the model to the new sample and repeat sets 2–3 multiple times. Boosting increases the weight when an observation is classified incorrectly. The key is to pay attention to your misclassified data points. Thus, you increase the weight age and keep doing this until all your misclassified samples are correctly predicted. In general, boosting decreases bias error and builds strong predictive models.

The common types of boosting are AdaBoost (Adaptive Boosting) and Gradient Boosting.

5.6.2.1 AdaBoost

AdaBoost or adaptive Boosting combines several weak learners into a single strong learner. It starts by assigning equal weights for all your data points and drawing a decision stump for a single input feature, so the next step, the results that you receive from the first decision stump, are then analyzed. If any observation is misclassified, then they are assigned a higher weight.

We use an iterative optimization process that is much more tractable. It adds the weak learners one by one, looking at each iteration for the best possible pair (coefficient, weak learner) to add to the current ensemble model. If any observations are misclassified, they are given a higher weight, and this process continues until all the observations fall into the right class.

Adaboost can be used for both classification and regression-based problems. However, it is more commonly used for classification purposes.

To illustrate how the algorithms works, consider a dataset $D = \{(x_1, y_1), ..., (x_m, y_m)\}$ where x_i is the input attribute and y_i is the output attribute}.

Now, for $t = 1,2,..., T$ (Iterations), we train a weak learner by using D_t and we get a weak classifier denoted by h_t. The aim is to select h_t with low weight error.

Hence, we choose α_t to minimize the training error such that:

$\alpha_t = \frac{1}{2} \ln[(1 - \epsilon_t/\epsilon_t)]$ where ϵ_t is the error of h_t and it is given by $\epsilon_t = \sum_{i=1}^{m} D_t(i)\delta h_t(x_i)$ with $h_t(x_i) \neq y_i$, is the sum of misclassifications.

For $i = 1, 2, ..,m$, we have $h_t(x_i) = y_i = \pm 1$ and $D_t(i)$ is the weight of (i), the training example where $D_1(i) = \dfrac{1}{m}$ for all (i).

Then, we train a weak learner by using D_t and we get a weak classifier denoted by h_t.

We choose α_t a real number, which is the weight associated with h_t and we update the distribution, which implies that: $D_{t+1}(i) = \left[D_t(i) e^{(-\alpha_t y_i h_t(x_i))} \right] / Z_t$ where Z_t is a normalization factor that is used to ensure that weights represent a true distribution. We have $\displaystyle\sum_{i=1}^{m} D_t(i) = 1$ where $Z_t = \displaystyle\sum_{i=1}^{m} D_t(i) e^{(-\alpha_t y_i h_t(x_i))}$ and $h_t(x_i) = 1$ or -1.

Now, the output is given by $H(x) = \sin\left(\displaystyle\sum_{t=1}^{T} \alpha_t h_t(x) \right)$ where if $\displaystyle\sum_{t=1}^{T} \alpha_t h_t(x))$ ispositive, then $H(x)$ is positive. Otherwise, it is negative.

5.6.2.2   Gradient Boosting

Gradient boosting is a robust machine learning algorithm that performs regression, classification, and ranking.

It is also based on the sequential ensemble model. This is where the base learners are generated sequentially so that the present base learner is always more effective than the previous one. The overall model improves sequentially with each iteration. The weights for misclassified outcomes are not incremented, or we will not add weights to the misclassified outcomes. Instead, we optimize the loss function of the previous learner by adding a new adaptive model that adds weak learners to reduce the loss function. The main idea is to overcome errors in the last learner's prediction.

It has three main components, the loss function that needs to be optimized, meaning that you need to reduce the error; the second component is the weak learners required for computing predictions and generate strong learners. Lastly, you will need an additive model that will regulate the loss function, which will fix the loss or the error from the previous weak learner.

Gradient Boosting requires less data pre-processing (than neural networks, but same as forests), often outperforms other classes of models, as boosting reduces the correlation of the trees' predictions, which improves the predictions of the boosting model. On the other hand, training is longer because trees are built sequentially. Also, Gradient boosting models are more sensitive to extreme values and anomalies in the data. They can be slow for real-time scoring.

References

Breiman L, Friedman J, Olshen R, Stone C (1984) Classification and regression trees. Chapman Hall/CRC (Orig. Published by Wadsworth), Boca Raton

Moro S, Cortez P, Rita P (2014) A data-driven approach to predict the success of bank tele-marketing. Decis Support Syst 62:22–31. Elsevier. https://archive.ics.uci.edu/ml/datasets/Bank+Marketing

Quinlan JR (1993) C4.5: programs for machine learning. Morgan Kaufmann, San Francisco

Yeh C (1998) Modeling of strength of high-performance concrete using artificial neural networks. Cem Concr Res 28(12):1797–1808. Concrete Compressive Strength Data Set. http://archive.ics.uci.edu/ml/datasets/Concrete%2BCompressive%2BStrength

Chapter 6
Regression Models

Learning Objectives

- Describe functions and their mathematical implementation
- Explain linear regression and multiple regression models
- Applications of regression models
- Explain and demonstrate logistic regression modeling

6.1 Functions and Mathematical Implementation

Many argue that data science can be done without knowing its math. Still, it is useful to look beneath the hood of some mathematical tools to understand and avoid critical errors in your analysis. This chapter states the following: types of functions, derivatives of functions, and matrices.

6.1.1 Functions

A function is like a machine; it has an input and output, and the outcome depending on the input. A function from a set A to a set B determines each element x (input) in A, a unique element y (output) in B, the image of x. Set A is the domain, and B is the image or range set of the function. A function is denoted by a letter f, g, h, ..., or letter of the Greek alphabet, such as sigma indicated by σ. For instance, $f(x)$ is the image of x under the function f.

© The Author(s), under exclusive license to Springer Nature
Switzerland AG 2022
L. Halawi et al., *Harnessing the Power of Analytics*,
https://doi.org/10.1007/978-3-030-89712-3_6

6.1.1.1 Types of Functions

There are different types of functions; some of them are listed in Table 6.1. However, this chapter covers the polynomial (linear), logarithmic, sigmoid, and exponential functions.

6.1.1.2 Linear Function

In Table 6.1, we stated some types of functions used from now on. A linear function is a polynomial function of degree 1, representing the regression line equation. A linear function has the form $y = a + bx$, where x is the independent variable, y is the dependent variable, and b is the linear function slope. The slope, b, is the ratio of the change in y to the change in x (a ratio of vertical change to horizontal change), and a is the Y-intercept (the value of Y when $x = 0$). The representation of a linear function is a straight line on a coordinate plane.

6.1.2 Coordinate Plane

Coordinate planes are essential to comprehend as they facilitate reading graphs, understanding points in space, and applying concepts in other subjects like data science and coding. The coordinate plane may be utilized to plot points, lines, and curves. This procedure allows us to describe algebraic relationships visually and aids us in building and interpreting algebraic concepts.

Interpreting the shape of a regression line, particularly the method of fitting in a coordinate plane and predicting the value of a dependent variable from an independent one, requires knowledge of the coordinate plane. A coordinate plane contains two perpendicular axes where the horizontal line is the x-axis, and the vertical line is the y-axis. These lines intersect at their zero points. This point is called the origin O. The position of a point relative to these axes is given by its coordinates (x, y). For example, $(2, 5)$ indicates this point's location by drawing a vertical line through

Table 6.1 Types of functions

Function	Example	Domain	Range
Polynomial	$f(x) = 2x^3 - 4x + 5$	R (the set of real numbers)	The set of real numbers
Exponential	$f(x) = e^x$	$(-\infty, +\infty)$	$(0, +\infty)$
Trigonometric	$f(x) = \sin x$	R (the set of real numbers)	$(-1, 1)$
Logarithmic	$f(x) = \log_e(x) = \mathrm{Ln}(x)$	R^+ (the set of positive real numbers)	R (the set of real numbers)
Sigmoid	$f(x) = \dfrac{1}{1+e^{-x}}$	R (the set of real numbers)	Between 0 and 1

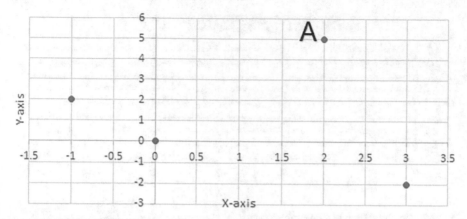

Fig. 6.1 Plot of A, O (0, 0), (−1, 2) and (3, −2)

$x = 2$ and a horizontal line through $y = 5$, then these two lines meet at point A. In Fig. 6.1, we plot the points A, O (0, 0), (−1, 2), and (3, −2).

6.1.2.1 Representation of a Linear Function in Graphical Form Example

Consider a phone company that charges a monthly service fee of $30 plus $0.6 per minute talk time. Let x equal the number of minutes consumed per month, and c is the total monthly fee. Then, the equation form, which represents this linear function, is $c = 0.6x + 30$. We can plot c on the coordinate plane (Fig. 6.2). Plotting a series of inputs and the calculated outputs gives an idea of the relationship between inputs and outputs of the function.

We can then use the representation of linear function to predict the value of independent variable c from the dependent variable x and vice versa.

For instance, as Fig. 6.2 shows, if we consume 10 min, we have to pay $c = 0.6.10 + 30 = \$36$, and if we have to pay $90, then $90 = 0.6x + 30$, which implies that $0.6x = 60$, and then $x = 100$.

Note: It would not be sensible to predict the value of c when $x > 500$ h from the above equation; we do not know whether the relationship will continue to be linear. The process of predicting a value from outside the range of your data is called extrapolation.

6.1.3 Derivative of Functions

The derivative of a function $y = f(x)$ is a measure of the rate at which the value y of the function changes with respect to the change of the variable x. We use differentiation in almost every part of machine learning, where the curve's direction depends on the slope or gradient.

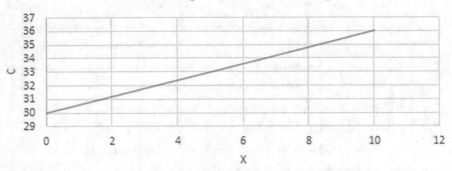

Fig. 6.2 Graphical form of a linear function

Suppose y is a function of x that $y = f(x)$. Then, $\dfrac{df(x)}{dx}$ is the derivative of y withrespect to x, and it is the gradient of the tangent to the curve $y = f(x)$. Optimization is at the heart of almost all machine learning and statistical techniques used in data science.

Optimization is a problem meant to maximize or minimize some function. Now, maximizing or minimizing function happens when the tangent is horizontal. Thatimplies the derivative $\dfrac{df(x)}{dx}$ is 0 at points x_0 at which $f(x_0)$ is a maximum or a minimum.

Table 6.2 represents some functions, their derivatives rules, and examples:

6.1.4 Matrices

Since we deal with a potentially large number of predictors in the multiple regression setting, it is more efficient to use matrices to outline the regression model and the subsequent analyses.

A matrix is used to summarize data; see Tables 6.4 and 6.5. Therefore, it is useful to define a matrix, a transpose matrix, and an inverse matrix, and how we operate with matrices are considered.

A matrix can store multiple values as an array of numbers with m rows and n columns. $(a_{ij})m \times n$ is the (i,j) the entry of matrix represents the ith row and jth column of a model, which is denoted by $A = (a_{ij})m \times n$. For instance, B is a 3×2 matrix with three rows and two columns where $a_{31} = 0$ is represented in Fig. 6.3.

Table 6.2 Derivative of functions

Functions	Derivative rules	Examples	Answers
$f(x) = ax^n$ where a is a constant, x is a variable and n is a real number	$\dfrac{df(x)}{dx} = nax^{n-1}$	$f(x) = 5x^{-0.5}$ where $a = 5$ and $n = -0.5$	$\dfrac{df(x)}{dx} = -0.5(5)x^{-0.5-1} = -2.5x^{-1.5}$
$f(x) = e^{ax}$ where a is a constant, x is a variable	$\dfrac{df(x)}{dx} = be^{ax}$	$f(x) = 2.5e^{-3x}$ where $b = 2.5$ and $a = -3$	$\dfrac{df(x)}{dx} = 2.5(-3)e^{-3x} = -7.5e^{-3x}$
$f(x) = \log(u(x))$ where $u(x)$ is a function of x	$\dfrac{df(x)}{dx} = \dfrac{\frac{du(x)}{dx}}{u(x)}$	$f(x) = \log(3x - 1)$	$\dfrac{df(x)}{dx} = \dfrac{3}{3x-1}$ where $\dfrac{du(x)}{dx} = 3$.
$h(x) = f(x) + g(x)$/summation of functions	$\dfrac{dh(x)}{dx} = \dfrac{df(x)}{dx} + \dfrac{dg(x)}{dx}$	$h(x) = 2.5x^4 - 2e^{5x}$	$\dfrac{dh(x)}{dx} = 2.5(4)x^{4-1} - 2.5e^{5x} = 10x^3 - 10e^{5x}$
$h(x) = f(x)g(x)$/product of functions	$\dfrac{df(x)}{dx} = \dfrac{df(x)}{dx}g(x) + f(x)\dfrac{dg(x)}{dx}$	$h(x) = e^{3x}(8x^{-4} - 10)$	$\dfrac{dh(x)}{dx} = 3e^{3x}(8x^{-4} - 10) + e^{3x}(-32x^{-5}) = e^{3x}$ $(24x^{-4} - 32x^{-5} - 30)$
$h(x) = \dfrac{f(x)}{g(x)}$ /quotient of functions	$\dfrac{dh(x)}{dx} = \dfrac{\frac{df(x)}{dx}g(x) - f(x)\frac{dg(x)}{dx}}{g(x)^2}$	$h(x) = \dfrac{1}{1+e^{-x}}$	$\dfrac{dh(x)}{dx} = \dfrac{0(1+e^{-x}) - 1(-1e^{-x})}{(1+e^{-x})^2} = \dfrac{e^{-x}}{(1+e^{-x})^2}$ where the derivative of a constant is zero

Fig. 6.3 Matrix

$$\begin{matrix} -1 & 2 \\ 5 & 3 \\ 0 & -2 \end{matrix}$$

Now, let A be a matrix with m rows and n columns. Then, the transpose of matrix A, denoted by A^T with n rows and m columns, switches the rows and columns. Now the inverse of B $n \times n$, denoted by B^{-1}, is a square matrix (number of rows equal the number of columns) where $B* B^{-1} = I$, where I is the identity matrix, a square matrix with 1 s on the main diagonal and 0 s everywhere else.

Let $A = (a_{ij})$ and $B = (b_{ij})$ be $m \times n$ matrices and let c be a number.

Then $A + B$ $(a_{ij} + b_{ij})m \times n$, the sum of A and B; $A - B = (a_{ij} - b_{ij})m \times n$, the difference of A and B, and $cA = (ca_{ij})m \times n$, the scalar multiple of A by c. Now, multiplying the matrices A and B is a different story. We should have the number of columns in A equal to the number of rows in B. Therefore, let A be a matrix with $m \times n$ and B a matrix with n rows and p columns. Then, $AB = A \times B$ is a matrix D with m rows and p columns with d_{ij} represents the ith row and jth column of the matrix D where $d_{ij} = \sum_{k=1} a_{ik} b_{kj} = a_{i1}b_{1j} + a_{i2}b_{2j} + \ldots + a_{im}b_{nj}$.

6.1.5 Definition of Logarithmic Function

To obtain a clearer picture of the calculation in Sect. 6.4, we show the relationship between an ln function and an exponential function and the rules and properties of Natural Logarithms.

The function $y = \mathrm{Ln}x$ is the inverse of the exponential function $y = e^x$ (Table 6.3). This inverse means:

- e^x lets us plug-in time and gets growth.
- Ln (x) enables us to plug-in growth and get the time it would take.

6.2 Linear Regression

Regression analysis is essential where it studies the relationships between data points and can help predict the near- and long-term business. Regression offers a different approach to prediction compared to decision trees. There are innumerable forms of regressions, which can be performed, such as linear, logistic, ridge, and lasso regression, among others.

Regressions, as parametric models, assume a specific association structure between inputs and targets. Linear regression is perhaps one of the most well-known

Table 6.3 Natural logarithm rules and properties

Rule name	Rule
Product	$\text{Ln}(x * y) = \text{Ln}(x) + \text{Ln}(y)$
Quotient	$\text{Ln}\left(\dfrac{x}{y}\right) = \text{Ln}(x) - \text{Ln}(y)$
Power	$\text{Ln}(y^x) = x\,\text{Ln}(y)$, $\text{Ln}(e^x) = x\text{Ln}(e) = x$ where $\text{Ln}(e) = 1$
Derivative	$\dfrac{\partial \text{Ln}(x)}{\partial x} = \dfrac{1}{x}$

Table 6.4 Linear regression notation

Observation number	Response Y	Explanatory variables $X_1 \quad X_2 \quad \ldots \quad X_K$
1	Y_1	$X_{11} \quad X_{12} \ldots X_{1k}$
2	Y_2	$X_{21} \quad X_{22} \ldots X_{2k}$
$\vdots$	$\vdots$	$\vdots$
n	Y_n	$X_{n1} \quad X_{n2} \quad X_{nk}$

Table 6.5 N equations

$$
\begin{bmatrix} Y_1 \\ Y_2 \\ \vdots \\ Y_n \end{bmatrix}
=
\begin{bmatrix} X_{11} & X_{12} \ldots & X_{1k} \\ X_{21} & X_{22} \ldots & X_{2k} \\ \vdots & \vdots & \vdots \\ X_{n1} & X_{n2} & X_{nk} \end{bmatrix}
\begin{bmatrix} \beta_0 \\ \beta_1 \\ \vdots \\ \beta_k \end{bmatrix}
+
\begin{bmatrix} \varepsilon_1 \\ \varepsilon_2 \\ \vdots \\ \varepsilon_k \end{bmatrix}
$$

and well-understood statistics and machine learning algorithms. Linear regression is a predictive algorithm that provides a linear relationship between the input X and the output or prediction Y, creating an equation. In simple linear regression, both the predictor variables and the target variable are assumed to be continuous, and it is assumed the dependent variable "Y" is contrasted with the independent variable(s) "X," and these independent observations are controlled.

Cases are scored using a simple mathematical prediction formula. Validation data allow the calculation of fit statistics. Values can be predicted within the range framed by the data, where this equation minimizes the distance between the fitted line and all the data points. The simple linear regression (SLR) model is identical to the slope-intercept form equation (Sect. 6.1.1.2).

The best fit (or trend line) line is an educated guess about where a linear equation might fall in a set of data plotted on a scatter plot (a coordinate plane). The equation in terms of each point is $y_i = \beta_0 + \beta_1 x_i$, where β_0 is a parameter, reflects the y-intercept, β_1 the regression slope, x_i is the regressor (independent variable), and y_i is the response variable (dependent variable).

The best fit line or the least square line is obtained by finding the values of β_0 and β_1 (denoted in the solutions as $\hat{\beta}_0$ and $\hat{\beta}_1$), whereby minimizing the sum of the squared vertical distances from all the points to the

line: $\Delta = \sum d_i^2 = \sum \left(y_i - \hat{Y}_i \right)^2 = \sum \left(y_i - \left(\beta_0 + \beta_1 x_i \right) \right)^2$.

Hence, the solutions are found by solving the equations where $\dfrac{\partial \Delta}{\partial \beta_0} = 0$ and $\dfrac{\partial \Delta}{\partial \beta_1} = 0$ are the partial derivatives of Δ (for more details about partial derivatives, refer to Chap. 7). That implies the least-squares (LS) estimation of β_0 and β_1 have the following formulas, where n is the number of observations.

$$\hat{\beta}_1 = \frac{n \sum_{i=1}^{n} x_i y_i - \sum_{i=1}^{n} x_i \sum_{i=1}^{n} y_i}{n \sum_{i=1}^{n} x_i^2 - \left(\sum_{i=1}^{n} x_i \right)^2} \quad \text{and} \quad \hat{\beta}_0 = \frac{1}{n} \left(\sum_{i=1}^{n} y_i - \hat{\beta}_1 \sum_{i=1}^{n} x_i \right)$$

Then, the equation of the fitted least-squares regression line is. $\hat{Y} = \beta_0 + \beta_1 x$. We can see that linear regression is extremely sensitive to outliers.

The correlation coefficient r on its own is very useful—the **correlation coefficient r**, can vary between -1.0 and 1.0. When the correlation (r) is negative, the regression slope (b) will be negative. When the correlation is positive, the regression slope will be positive.

Note: The coefficient of determination is also called r-square and is denoted as r^2. The coefficient of determination is the portion of the total variation in the dependent variable explained by variation in the independent variable.

The regression slope β_1 also is called the correlation coefficient, which indicates the strength and the sign of the relationship between two variables. For instance, if $\beta_1 < 0$, then the correlation is negative, and if $\beta_1 > 0$, then the correlation is positive. If the coefficient approaches zero, then we do not have a linear relationship between the two variables.

6.2.1 Applications of Linear Regression

We have our data series with multiple observations and variables. We have covered descriptive statistics (central tendencies, variance, standard deviation) in Chap. 3, and we want to move into a more sophisticated analysis with our data by building simple models.

The critical question that analysts want to address is how these variables in the data set are related. Correlations are useful to indicate possible relationships between variables.

Further, one or more variables can often be employed to predict another variable. For example, analysts can ask about the relationship between people's height and weight, ethnicity, healthcare options, yearly income, and level of education achieved. Another example using the aviation data set, the analyst may be looking at the impact of distance on total plane time. How much does distance impact flight time? How are arrival delays impacted by day of the week, distance group, and arrival time block. Any one or combination of variables? How is total plane time impacted by distance or day of the week?

We can examine the prediction of airport arrival rates based on weather factors and other available a priori input variables. With an airport's response to various weather conditions better understood, arrival rates could be objectively estimated with greater skill (perhaps out to several days) using predictive numerical weather guidance. The analysts may be interested in predicting how much distance impacts total plane time or flight time.

First, can regression be used to discover significant relationships between various weather and delays variable inputs and airport efficiencies? Second, what factors can then be used as inputs to estimate arrival delays?

This is called a model or a regression equation. Sometimes the term regression line is used interchangeably with the term regression equation. We introduced linear models in Sect. 6.1.2.1.

6.3 Multiple Linear Regression

Assuming that more than one factor influences the response, we have a dependent variable y and two or more independent variables. Therefore, multiple regression describes how a response variable y depends linearly on many predictor variables. Thus, the population regression line for exploratory variables is $(x_1, x_2, ..., x_k)$, where k is a natural number. For instance, let y be the selling price of a house. Then, y depends on the following factors (location, number of bathrooms, number of bedrooms, the year the house was built, garden, city, ...).

Multiple regression describes how a response variable y depends linearly on many predictor variables. Regressions predict cases using a mathematical equation that involves the values of the input variables. The multiple linear regression model is denoted by

$$y = \beta_0 + \beta_1 x_1 + \beta_2 x_2 + \cdots \beta_k x_k + \varepsilon.$$

The parameters $\beta_0, \beta_1, \ldots, \beta_k$ are the regression coefficients associated with $x_1, x_2, \ldots, x_k$ respectively, and ε is the random error component reflecting the difference between the observed and fitted linear relationship.

Multiple regression is appropriate for the analysis of experimental or nonexperimental research. We can go with forward selection, backward elimination, stepwise approach, and Lasso to select the most significant independent variables.

The regression node in SAS Viya provides four sequential selections, forward, backward, stepwise, and Lasso.

- Forward selection creates a sequence of models of increasing complexity. We start with no variables and add predictors with a p value less than the critical value. The algorithm searches the set of one-input models and selects the model that most improves the baseline model.
- Backward selection creates a sequence of models of decreasing complexity. The sequence starts with all the variables and then removes the predictor with the highest p value fit statistic. The input chosen for removal reduces the overall model fit statistic at each step. This is equivalent to removing the input with the highest p value. The least significant effects are removed one at a time until the model is significantly weakened by removing an effect.
- Stepwise selection is a combination of backward elimination and forward selection. The process terminates when all inputs available for inclusion in the model have p values over the entry cutoff. All inputs already included in the model have p values below the stay cutoff.
- The lasso method adds and removes candidate effects based on a version of ordinary least squares, where the sum of the absolute regression coefficients is constrained. Modern regression techniques (e.g., "ridge and lasso" regression) "are very useful" when the number of variables exceeds the number of observations or when collinearity is suspected between the predictor variables.

There are many advantages of multiple regression (MR). MR can use both categorical and continuous independent variables to include multiple independent variables.

There are also some disadvantages associated with multiple regression. For instance, the dependent variables must be a linear function of the independent variables; each observation should be independently drawn from the population and the associated errors. Errors should be normally distributed. The variance of errors should not be a function of the independent variables. The dispersion of values along the regression line should be fairly constant for all values of X (homoscedasticity).

6.3.1   Estimation of the Model Parameters

The running of any machine-learning algorithm is portrayed as a function $f(x)$. The function f is established during the training phase (aka the learning phase). Usually, this stage includes using a set of known data pairs $\{x, y\}$ to estimate f. Once the model is trained, it predict on a test set for x that it has not seen in the training phase. Usually, the function $f(x)$ will have parameters that need to be tuned for optimal performance.

Intercept and parameter estimates are chosen to minimize the squared error between the predicted and observed target values (least squares estimation). The prediction estimates can be viewed as a linear approximation to the expected (average) value of a target conditioned on observed input values.

Now, suppose that we have n observations on the $k + 1$ variables where $Y_i = \beta_0 + \beta_1 x_{i1} + \ldots + \beta_k x_{ik} + \varepsilon_i$ with $i = 1, \ldots, n$ and n should be more significant than k. We can think of the observations as points in $(k + 1)$-dimensional space. Hence, our goal in the least-squares regression is to fit a hyperplane into $(k + 1)$ dimensional space that minimizes the sum of squared residuals. Whereby finding the value of these

parameters $\beta_1, \beta_2, \ldots, \beta_k$ that happens by minimizing $\sum \varepsilon_i^2 = \sum_{i=1}^{n} \left(y_i - \beta_0 - \sum_{j=1}^{k} \beta_j x_{ij} \right)^2$.

For multiple regression, we have to calculate $k + 1$ partial derivatives of $\sum \varepsilon_i^2$ concerning the model parameters $\beta_0, \ldots, \beta_k$, set them equal to zero, and derive the least-squares normal equations that our parameter estimates $\beta_0, \beta_1, \ldots, \beta_k$.

Consider we work in $(k + 1)$-dimensional Euclidean space, then solving these equations is much more conveniently formulated with the help of vectors and matrices.

The linear regression model can be written in the form $Y = X\beta + \mathcal{E}$, with the following compact notation, as presented in Table 6.4.

Then, the n equations can be written as follows (Table 6.5)

Now, let $S(\beta) = \sum \varepsilon_i^2$. We assume that $S(\beta)$ is a real-valued, convex, differentiable function to ensure that its minimum exists. Therefore, $\sum \varepsilon_i^2 = \varepsilon^T \varepsilon$, where $\mathcal{E}^T$ is the transpose of $\mathcal{E}$. Then, $\varepsilon^T \varepsilon = (Y - X\beta)^T (Y - X\beta) = Y^T Y + \beta^T X^T X \beta - 2\beta^T X^T Y$. Now,

differentiate $S(\beta)$ concerning β, we have $\dfrac{\partial S(\beta)}{\partial \beta} = 2X^T X \beta - 2X^T Y$ (note that if Z is an $m \times 1$ vector and A is any $m \times m$ symmetric matrix). We have $Z^T AZ$,

then $\dfrac{\partial Z^T AZ}{\partial Z} = \left(A + A^T \right) Z = 2AZ$.

Hence, we apply this result in the above). Then, we minimize $S(\beta)$ by setting $\dfrac{\partial S(\beta)}{\partial \beta} = 0$, which implies that $2X^T X \beta - 2X^T Y = 0$. Then, we have $X^T X \beta = X^T Y$

where $\beta = (X^T X)^{-1} X^T Y$, termed as ordinary least squares estimator (OLSE). Since $\dfrac{\partial^2 S(\beta)}{\partial \beta^2}$ is at least nonnegative definite, so β minimizes $S(\beta)$.

The squared multiple correlation coefficient is r^2, representing the total variance in the dependent variable with respect to the independent variables.

$$r = \frac{n\left(\sum xy\right)-\left(\sum x \sum y\right)}{\sqrt{\left[n\sum x^2 -\left(\sum x\right)^2\right]\left[n\sum y^2 -\left(\sum y\right)^2\right]}}.$$

- r = The correlation coefficient
- n = number in the given dataset
- x = first variable in the context
- y = second variable

Adjusted R square or modified R^2 is a modified version of R squared, it adjusts for predictors that are not significant in the model.

It does not consider the impact of all independent variables but only those that impact the variation of the dependent variable. The value of the modified R^2 can be negative also, though it is not negative most of the time. A higher value indicates that the additional input variables are adding value to the model.

$$r_{adj}^{2} = 1 - \left[\left(1-r^2\right)\left(\frac{n-1}{n-k-1}\right)\right]$$

(where n = sample size, k = number of independent variables)

- Penalizes excessive use of unimportant independent variables.
- Smaller than r^2
- Useful in comparing among models

The p value is an essential statistical measure and one of the best ways to test and validate if the results are statistically significant. P value is computed using the statistical hypothesis testing technique. It is the probability that a given statistical model null hypothesis is true. The higher the p value, the more likely our null hypothesis is true, and the more likely variable X does not contribute to concrete compressive strength.

6.3.2   Concrete Compressive Strength Example 1: SAS Visual Analytics

The data set used in this example comprises 1030 instances with nine attributes consisting of eight quantitative input variables and one quantitative output variable (Yeh 1998). We do not have any missing values. Concrete is an important material in civil engineering. The concrete compressive strength is a highly nonlinear function of age and ingredients, and we will use it as our target variable (refer to Appendix D)

For this demonstration, we are looking at answering the following question: What determines the compressive strength of concrete?

The best practice before creating any model is to explore the data set and see the characteristics of the variable as presented in Table 6.6. Table 6.7 presents the model fit statistics.

Once we explored our data, a second step would include dealing with missing data, thus the impute note. We did not have any missing variables in our data set, so the impute node did not make any significant difference to the data set.

Using SAS Visual Analytics, we can create an interactive regression model using the stepwise method as presented in Fig. 6.4.

For linear regression, the evaluation criteria choices are as follows:

- Adjusted R-square
- AIC
- AICC
- Average squared error
- F value for model
- Mean square error
- R-square
- Root MSE
- SBC
- Validation ASE

The top of the linear object model highlights the target variable chosen (concrete compressive strength), the overall fit statistics (ASE and the number of observations used by the model). In this case, 1030 observations were used.

SAS VIYA supports several common fit statistics when building linear regression models, including AIC, R square, ASE, and SFBC. In our model, we focus on the validation ASE (average square error). ASE is calculated as the sum of squared errors divided by the number of observations; it highlights the average difference (error) between the actual value (compressive strength) and the predicted value across all the data points.

Our model has a validation ASE of 109.01, meaning that, on average, our model has produced predictions that are off from the actual value by 109.01. One of the goals of our model tuning steps would be to reduce this value to predict the compressive strength with a higher degree of accuracy and produce a better fit model.

The fit summary window lists all the variables used in the model ranked by their p values, which are used to determine the significance of the variables. Our example in the fit summary graph, shows that age, cement component 1, blast furnace, fly ash, and water have the longest bars, meaning they have small p values and are important factors when predicting compressive strength.

The residual plot shows the relationship between the predictive value (concrete compressive strength) and the residual of that observation. The predictions of the values are shown on the X-axis, and the accuracy of the predictions is shown on the

Table 6.6 Descriptive statistics concrete data set

Output

Input variable statistics

Obs	Input variable	Measurement level	Number of missing values	Percentage missing	Imputable	Minimum	Maximum	Mean	Midrange	Standard deviation	Skewness	Kurtosis	Label
								–		–	–	–	–
1	Age (day)	Nominal	0	0	0	0	359.4	73.676488673	179.700	86.546278997	0.8056400554	−0.525235294	
2	Blast furnace slag (component 2)	Interval	0	0	0	0	540	282.20571197	321.000	104.62491427	0.5230849954	−0.47456494	
3	Cement (component 1)	Interval	0	0	0	102	1145	972.43606796	973.000	78.945726202	−0.030581824	−0.61130916	
4	Coarse Aggregate (component 6)	Interval	0	0	0	801	992.6	773.11917476	793.300	78.83051885	−0.308532358	−0.201997739	
5	Fine aggregate (component 7)	Interval	0	0	0	594	200.1	53.470776699	100.050	62.890097777	0.5303099688	−1.334445242	
6	Fly ash (component 3)	Interval	0	0	0	0	32.2	6.2757831715	16.100	6.0003793436	0.9549812127	1.6831563985	
7	Superplasticizer (component 5)	Interval	0	0	0	0	237	181.76864078	179.375	21.023983328	0.0291092844	0.0208738485	
8	Water (component 4)	Interval	0	0	0	121.75							

Table 6.7 Model fit statistic

Linear regression—cement (component 1) 1 supplement 2						
Source	Deg freedom	Sum of squares	Mean square	F value	Pr > F	R-square
Model	7	5,624,526	803,503.6	643.3316	<0.00001	0.880704
Error	610	761,873.4	1248.973	–	–	–
Corrected total	617	6,386,399	–	–	–	–

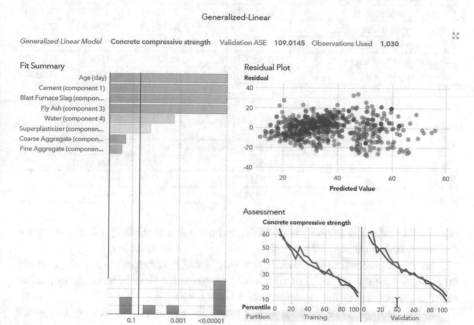

Fig. 6.4 Results of regression model

Y-axis. The vertical distance between the actual predictions and the horizontal like at ZERO at any given predicted value shows the residual value and how bad the prediction is for that value. Positive values above 0 mean that the prediction was too low, and negative values below 0 on the Y-axis mean that the prediction was too high. O means the prediction was exactly correct.

When we expand the linear regression object, a table is displayed with additional metrics associated with the regression model. The assessment plot reflects the performance characteristics from the model we selected.

The fit statistics include all the model fit statistics in tables that are useful as it enables us to assess other important model fit statistics such as R-square. The correlation coefficient r provides us with both the strength and direction of the linear relationship between the independent variables and our dependent variable—compressive strength—the values of r range between −1 and +1. In our example, the

Table 6.8 Parameters estimates

Parameter	Estimate	Standard error	t valve	Pr > [t]
Linear regression—concrete compressive strength 1 supplement 4				
Intercept	−3.88535	33.79974	−0.11495	0.90852
Coarse aggregate (component 6)	0.006961	0.012127	0.573997	0.56618
Age (day)	0.115259	0.007284	15.82419	<0.00001
Blast furnace slag (component 2)	0.092376	0.013065	7.070638	<0.00001
Cement (component 1)	0.114925	0.010898	10.54542	<0.00001
Fine aggregate (component 7)	0.010995	0.013596	0.808708	0.41900
Fly ash (component 3)	0.087424	0.016248	5.380423	<0.00001
Superplasticizer (component 5)	0.326039	0.123249	2.645362	0.00837
Water (component 4)	−0.14701	0.050597	−2.90561	0.00380

Table 6.9 Model fit

Partition	ASE	Observations used
Linear regression—cement (component 1) 1 (1) supplement 9		
Training	1284.7960	618
Validation	1139.7298	412

relationship between xs and y is positive, and the value is 0.8807, which means we do have a strong relationship.

The parameter estimates Table 6.8 displays the parameter estimates, standard error, t-value, and p value, all of which help us assess the significance and impact of the input variables for the model we are building. Table 6.8.

We can see age, blast furnace slag, cement fly ash, superplasticizer, and water variables have low p values and are significant.

In addition, the parameter estimates provide values or coefficients for the individual input variable. These parameters describe the relationship between the predictor and the target variables (concrete compressive strength). The sign of each parameter estimate indicates the direction of these relationships. This parameter estimates values represent the relative change in the target variable given a one-unit change in the predictor variable.

In our case, we can see that water has a negative estimate value (−0.14701) and, therefore, a negative relationship with compressive strength. This means that compressive strength would decrease as we improve or increase water. Furthermore, we should see an increase of (0.114925) in compressive strength for each cement increase.

The output of the linear regression model is essentially a mathematical formula. This formula provides a complete view of how the various predictor variables contribute to the target variable and interprets and communicates the model.

So if we take the most significant values a per the p value, we can describe our linear regression

With the following mathematical function (refer to Sect. 6.3):

$y = \beta_0 + \beta_1 x_1 + \beta_2 x_2 + \cdots \beta_k x_k + \varepsilon$ where y is compressive strength and $x_1, x_2, \ldots, x_6$ are the exploratory variables with $k = 7$.

Hence, $y = -3.88535 + 0.006961$ Coarse Aggregate 1 + 0.115259 Age − 0.14701 Water

$b_1 = -0.006961$: compressive strength will increase on average by 0.006961 for each unit increase in coarse aggregate 1, net of the effects of changes due to other factors.

In Table 6.9, we review the fit statistics for our model. We can see that the results improved with our validation data set.

6.3.3 Concrete Compressive Strength Example 1: Model Builder

In Model Studio, you can create a logical process flow in the form of a pipeline, as shown in Fig. 6.5. After creating a new pipeline, you can create visual data mining functionality by adding nodes to the pipeline.

For this demonstration, we created five different regression models with different selection methods. The model comparison node would identify the best model or the champion model.

For the concrete data set, the champion model was the regression model with the backward selection method, as presented in Table 6.10. We present the results for the Backward Selection. Just to recap, in this selection method, all candidate effects are included in the initial model. The least significant effects are removed one at a time until the model is significantly weakened by removing an effect. We run the pipeline. The results are highlighted in Tables 6.11 and 6.12.

The overall fit's statistical significance is determined with an F test by comparing the regression model variance to the error variance. The R-square estimate is an indicator of how well the model fits the data. (E.g., an R square close to 1.0 indicates that the model accounted for almost all the variability in the response variable with the predictor variables specified in the model.)

R square (adjusted) is recommended for model selection because the sample size and the number of predictors are used when the R-square estimate is adjusted.

The F test (Table 6.11) shows the overall significance of the model. It shows if there is a linear relationship between all of the X variables considered together and Y. The R square is 0.82456; it is solid and positive.

The overall ANOVA tab provides analysis-of-variance results for the model, error, and corrected total. The p value ($\Pr > F$) and R-square indicate the significance of individual inputs and overall model fit, respectively, as presented in Table 6.11. Adjusted R square indicates that 0.82% of the variation in the strength of concrete can be explained by the relationship between the concrete compressive strength and its predictors. It appears the model explains 82% of the variability in the data.

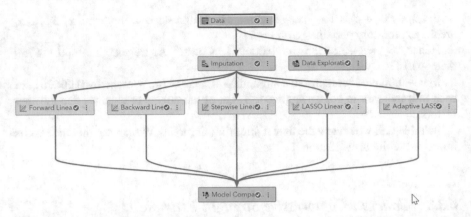

Fig. 6.5 Regression models—model studio

Table 6.10 Model comparison

Champion	Name	Algorithm name	Average squared error
	Stepwise linear regression	Linear regression	45.2396
X	Backward linear regression (1)	Linear regression	45.2396
	Forward linear regression	Linear regression	45.3968
	Adaptive LASSO	Linear regression	45.3968
	LASSO linear regression	Linear regression	45.6421

Table 6.11 Backward model fit statistics

The SAS system					
The REGSELECT procedure					
Selected model					
Analysis of variance					
Source	DF	Sum of squares	Mean square	*F* value	Pr > *F*
Model	16	139,114	8694.63005	176.54	<.0001
Error	601	29,599	49.24992		
Corrected total	617	168,713			
Root MSE			7.01783		
R-square			0.82456		
Adj *R*-Sq			0.81989		
AIC			3045.05083		
AICC			3046.19274		
SBC			2500.30114		
ASE (train)			47.89515		
ASE (validate)			45.23955		

Regression complexity is optimized by choosing the optimal model in the sequential selection sequence. The process involves two steps. First, fit statistics are calculated for the models generated in each selection process steps. Both the training and validation data sets are used. Then the simplest model (i.e., the one with the fewest inputs) with the optimal fit statistic is selected.

Fit statistics about the model results are included for training and validation data sets for each optimization process. Table 6.12 presents the fit statistics for the model comparison node. The mean square error is 0.0711 for our data validation set. Fortunately, regression models lend themselves to straightforward interpretation. The lower the value, the better, and 0 means the model is perfect.

6.4 Logistic Regression

Logistic regression extends the ideas of linear regression to the situation where the outcome variable is categorical rather than continuous. Logistic regression is a predictive model used to describe the relationship between one dependent binary variable and one or more ordinal, nominal, interval, or ratio-level independent variables. Logistic regression is used to find the probability of event = Success and event = Failure. We should use logistic regression when the dependent variable is binary (0/1, True/False, Yes/No) in nature.

Logistic regression is used in the biological sciences in many social applications, where instead of precisely predicting what is happening, you predict the probability or likelihood. For instance, to predict whether an email is spam (1) or (0) or whether the tumor is malignant (1) or not (0). In a business setting, logistic regression may be used for profiling, classification, and in any situation where you are describing choice behavior.

Logistic regression is widely used for classification problems. It does not require a linear relationship between dependent and independent variables. It can handle various types of relationships because it applies a nonlinear log transformation to the predicted odds ratio. To avoid overfitting and underfitting, we should include all significant variables. It requires large sample sizes because maximum likelihood estimates are less powerful at low sample sizes than ordinary least squares. The independent variables should not be correlated with each other, such as no multicollinearity.

How could we model and analyze such data? We use the conditional distribution of the response Y, given the input variables, Pr $(Y|X)$ (We know that probability takes a value between 0 and 1). Hence, assume that we have a binary output variable Y. We want to model the conditional probability $\Pr(Y = 1|X = x)$ as a function of x; any unknown parameters in the function are estimated by maximum likelihood. The statisticians approached this problem by asking themselves, "How can we use linear regression to solve this?"

Table 6.12 Fit statistics

Target name	Data role	Partition in...	Formatted...	Sum of fre...	Average sq...	Divisor for...	Root avera...	Mean abso...	Root mean...	Mean squa...
Concrete compressive strength	Validate	0	0	412	45.2396	412	6.7260	5.2854	2.2990	0.0711
Concrete compressive strength	Train	1	1	618	47.8952	618	6.9206	5.3267	2.3080	0.0517

The most straightforward modification of $\ln p$, which has an unlimited range, is the logistic (or logit) transformation, $\ln\left(\dfrac{p}{1-p}\right)$. The scientific approach is the following:

We know that $Y = \beta_0 + x\,\beta_1$. Therefore, if we apply the sigmoid function $p = \dfrac{1}{1+e^{-y}}$, then we solve for y from $\ln\left(\dfrac{p}{1-p}\right)$, we have

$$\ln\left(\frac{p}{1-p}\right) = Ln\left(\frac{\dfrac{1}{1+e^{-y}}}{1+\dfrac{1}{1+e^{-y}}}\right) = Ln\left(\frac{\dfrac{1}{1+e^{-y}}}{\dfrac{e^{-y}}{1+e^{-y}}}\right) = Ln\left(\frac{1}{1+e^{-y}}*\frac{1+e^{-y}}{e^{-y}}\right) = Ln\left(e^{y}\right) = y$$

We have $y = \beta_0 + x\,\beta_1$. Therefore, we can see that $\ln\left(\dfrac{p}{1-p}\right) = y = \beta_0 + x\beta_1$.

Hence, we develop this line, similar to the linear regression line, but it looks different. The logistic regression's main goal is to find the best fit line, as shown in Fig. 6.6.

We make this a linear function of x without fear of nonsensical results. Hence, formally, the logistic regression model is that

$$Ln\left(\frac{p(x)}{1-p(x)}\right) = \beta_0 + x\beta_1.$$

Note: We should predict $Y = 1$ when $p \geq 0.5$ and $Y = 0$ when $p < 0.5$ to minimize the classification rate, which means guessing 1 whenever $\beta_0 + x \cdot \beta$ is nonnegative, and 0 otherwise.

The main difference between linear and logistic regression is the structure of the dependent variable. For example, Linear regression assumes a quantitative numeric variable, while logistic regression assumes a qualitative categorical target variable.

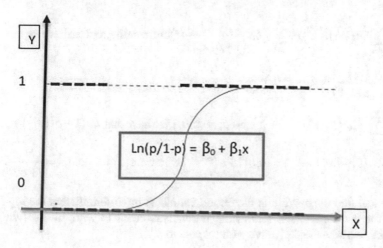

Fig. 6.6 Logistic regression best fit line

6.4.1 Estimating the Coefficients

Parameter estimates are obtained by maximum likelihood estimation.

Logistic regression does not have the same concept of a "residual," so it cannot use least squares; instead, it uses something called "maximum likelihood." Because logistic regression predicts probabilities, rather than just classes, we can fit it using likelihood.

Hence, with the maximum likelihood method, for each training data set (x_i, Y_i) with $1 < = i < = n$ and the probability of that class was either p, $y_i = 1$, or $1 - p$, if $y_i = 0$. The likelihood is the

$$L(\beta_0,\beta_1) = \prod_{i=1}^{n} p(x_i)^{y_i} \left(1-p(x_i)^{1-y_i}\right).$$

Then, the log-likelihood turns products into sums. In this case, it is easiest to use a base of e for the log of the likelihood, or natural log, ln, which equals $\log_e$; we have $\ln(e) = 1$. This makes the exponential part much easier to understand. Here are the steps for expressing the new log-likelihood function,

$$Ln(\beta_0,\beta_1) = Ln\prod_{i=1}^{n} p(x_i)^{y_i} \left(1-p(x_i)^{1-y_i}\right)$$

$$= \sum_{i=1}^{n}\left[Ln\, p(x_i)^{y_i} + Ln\left(1-p(x_i)\right)^{1-y_i}\right](\text{Product rule})$$

$$= \sum_{i=1}^{n}\left[y_i Ln\, p(x_i) + (1-y_i)Ln\left(1-p(x_i)\right)\right](\text{Power rule})$$

$$= \sum_{i=1}^{n}Ln\left(1-p(x_i)\right) + \sum_{i=1}^{n}y_i Ln\left(p(x_i)\right) - \sum_{i=1}^{n}y_i\, Ln\left(1-p(x_i)\right)(\text{Rearrangement})$$

$$= \sum_{i=1}^{n}Ln\left(1-p(x_i)\right) + \sum_{i=1}^{n}y_i Ln\frac{p(x_i)}{\left(1-p(x_i)\right)}(\text{Quotient rule, and we have ln}$$

$$\left(\frac{p(x_i)}{1-p(x_i)}\right) = \beta_0 + x_i\,\beta_1)$$

$$= \sum_{i=1}^{n}-Ln\left(1+e^{\beta_0+x_i\beta_1}\right) + \sum_{i=1}^{n} y_i\left(\beta_0 + x_i\,\beta_1\right)\text{We have that } Ln\left(1-p\left((x_i)\right)\right)$$

$$= Ln(1) - Ln\left(1+e^{y_i}\right) = -Ln\left(1+e^{\beta_0+x_i\beta_1}\right)\text{where } Ln(1) = 0)\text{and } p = \frac{1}{1+e^{-y}}.$$

We can find the maximum likelihood estimates by differentiating the log-likelihood for the parameters, setting the derivatives equal to zero, and solving. For instance, we take the derivative with respect to β_1.

We have $\dfrac{\partial Ln\left(1+e^{\beta_0+x_i\beta_1}\right)}{\partial\beta_1}=\dfrac{x_i e^{\beta_0+x_i\beta_1}}{1+e^{\beta_0+x_i\beta_1}}$ and $\dfrac{\partial\left(y_i\left(\beta_0+x_i\beta_1\right)\right)}{\partial\beta_1}=y_i x_i$. Then,

$$\frac{\partial Ln\left(\beta_0,\beta_1\right)}{\partial\beta_1}=-\sum_{i=1}^{n}\frac{e^{\beta_0+x_i\beta_1}}{1+e^{\beta_0+x_i\beta_1}}x_i+\sum_{i=1}^{n}y_i x_i.$$

Now, $\dfrac{\partial Ln\left(1+e^{\beta_0+x_i\beta_1}\right)}{\partial\beta_0}=\dfrac{e^{\beta_0+x_i\beta_1}}{1+e^{\beta_0+x_i\beta_1}}$ and $\dfrac{\partial\left(y_i\left(\beta_0+x_i\beta_1\right)\right)}{\partial\beta_0}=y_i$. Then,

$$\frac{\partial Ln\left(\beta_0,\beta_1\right)}{\partial\beta_0}=-\sum_{i=1}^{n}\frac{e^{\beta_0+x_i\beta_1}}{1+e^{\beta_0+x_i\beta_1}}+\sum_{i=1}^{n}y_i.$$

The maximum likelihood estimates for β_0 and β_1 can be found by solving the two equations, which are the results of setting $\dfrac{\partial Ln\left(\beta_0,\beta_1\right)}{\partial\beta_1}=0$ and $\dfrac{\partial Ln\left(\beta_0,\beta_1\right)}{\partial\beta_0}=0$.

Then, we can find the approximate numerical solutions.

It is important to note that we can use multiple logistic regression when we have one nominal and two or more measurement variables. Then, the maximum likelihood estimates for β, which can be found by setting each of the K equations equal to zero and solving for each β_k we have, then

$$\frac{\partial Ln\left(\beta_0,\ldots,\beta_k\right)}{\partial\beta_k}=-\sum_{i=1}^{n}\frac{e^{\sum_{k=0}^{K}x_{ik}\beta_k}}{1+e^{\sum_{k=0}^{K}x_{ik}\beta_k}}x_{ik}+\sum_{i=1}^{n}y_i x_{ik}=0,\text{ where }K\text{ is the number}$$

of independent variables specified in the model. Then, we can find the values of β_1, β_2, ..., β_k.

6.4.2   Types of Logistic Regression

There are additional types of logistic regression model summaries mentioned subsequently. However, their explanation is beyond the scope of the book.

1. Binary logistic regression—The categorical response has only two possible outcomes. For instance, Spam or Not
2. Multinomial logistic regression—Three or more categories without ordering. For example, predicting which food is preferred more (Veg, Non-Veg, Vegan)
3. Ordinal logistic regression—Three or more categories with this form. For instance, software rating from 1 to 4.

6.4.3   Demonstrations of Logistic Regression: Aviation Example

This section demonstrates a classification example using the aviation data set. We will use SAS visual statistics to build a logistics regression model to analyze arrival delays.

We will use the model to explore what drives arrival delays. First, we create a new logistics regression visualization using the backward method as presented in Fig. 6.4 to see what drives delays. Our target variable, Airline Delays (1 = delays, 0 no delay), is a binary variable indicating whether or not we have delays.

Next, we will select certain attributes that we will add to the analysis; these attributes are mainly related to delays, for example, day of the week, distance, departure block, arrival block, departure delays. Also, we want to know whether certain regions have more delays than others, so we add the origin city to the model as a classification effect.

The model is completed; 690 observations were used for our training data set and 461 observations for our validation data set.

Let us review the output now from Fig. 6.7.

The top of the logistic object model highlights the target variable is chosen (arrival delays log), the overall fit statistics KS (Youden). Our model has a 0.9336 KS (Youden). The maximum value of the Youden index is 1 (perfect test), and the minimum is 0 when the test has no diagnostic value. Its value ranges from 0 through 1 (inclusive). A value of 1 indicates no false positives or false negatives, that is, the test is perfect.

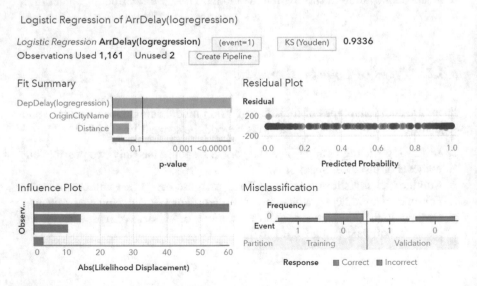

Fig. 6.7 Results of the regression model

The first visualization on the left is the variable importance plot. The influence plot shows the outlying leverage and influence of each case. This plot displays the negative log values of each variable's p value. Green indicates that the predictor is significantly related to the target variable (arrival delays), blue indicates insignificant predictors with a p value greater than 0.05. Our example in the fit summary graph shows that the dep delay log is the most significant variable and that the original city name and distance are insignificant.

The output includes a residual plot for model diagnosis. The Residual Plot window enables you to quickly identify potential outliers in the analysis. You can right click the statistics name to switch to other residuals such as deviance, Pearson residual, and standardized Pearson residual. It displays the relationship between the predicted data and the residual data using a scatter plot or a heat map to identify outlier observations. The residual plot displays the X-axis response variable's predicted value and the studentized, deleted residual statistic on the Y-axis.

If we like to further with our analysis, SAS visual statistics provides several model assessment visualizations such as the Lift chart, the ROC curve, or the misclassification chart.

The misclassification chart enables you to evaluate the quality of your model because you can see how many observations in the model were correctly or incorrectly classified for each response variable's value as presented in Fig. 6.8.

For this data, for the bar corresponding to the event level of y, "yes," the segment of the bar colored as "Correct" corresponds to true positives. The misclassification plot displays how many observations were correctly and incorrectly classified for each response variable's value. Based on the misclassification rate shown here (Fig. 6.9), it looks like we have a good model.

Logistic regression - ArrDelay(logregression) 1 Supplement 13

Response	Event	Value	Training Frequency	Validation Frequency
Correct	1	True Positive	95	69
Incorrect	1	False Negative	114	86
Correct	0	True Negative	465	281
Incorrect	0	False Positive	16	25

Fig. 6.8 Event classification plot

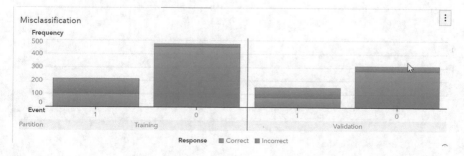

Fig. 6.9 Misclassification plot

Fit Statistics											
Target Name	Data Role	Partition In...	Formatted ...	Sum of Fre...	Average Sq...	Divisor for...	Root Avera...	Misclas...	Multi-Cl..	KS(You...	Area
ArrDelay (logregression)	TRAIN	1	1	698	0.0007	698	0.0268	0.0014	0.0000	1	
ArrDelay (logregression)	VALIDATE	0	0	465	0.0011	465	0.0328	0.0022	0.0000	1	

Fig. 6.10 Fit statistics

   Figure 6.10 displays the fit statistics. We can use the average squared error value or the KS to see how well our model fits. K-S should be a high value (Max = 1.0); in this case, we have an acceptable fit (KS is 1). Also, the ASE is 0.0011 for the validation data set; the lower the value, the better, and 0 means the model is perfect.

Reference

Yeh C (1998) Modeling of strength of high performance concrete using artificial neural networks. Cem Concr Res 28(12):1797–1808. Concrete Compressive Strength Data Set. http://archive. ics.uci.edu/ml/datasets/Concrete%2BCompressive%2BStrength

Websites

https://machinelearningmastery.com/linear-regression-for-machine-learning/
http://www.stat.cmu.edu/~cshalizi/uADA/12/lectures/ch12.pdf
https://www.analyticsvidhya.com/blog/2015/08/comprehensive-guide-regression/
https://machinelearningmastery.com/logistic-regression-for-machine-learning/
https://towardsdatascience.com/introduction-to-machine-learning-algorithms-linear-regression-14c4e325882a
https://statisticsbyjim.com/regression/predictions-regression/

Chapter 7
Neural Networks

Learning Objectives

- Define neural networks and how neural networks learn
- Identify the architecture and different terms used with neural networks
- Explain the math behind neural networks
- Develop a neural network model

7.1 What Are Neural Networks?

Neural networks, also called neural nets, have the smallest operational applications compared to supervised machine learning techniques. These applications are seen in financial applications such as bankruptcy prediction, commodity trading, and detecting fraud in credit card and monetary transactions. They were developed to imitate the function of the human brain. Neural networks are more flexible than traditional nonlinear models and can be used for classification or prediction. No function form is required, nor is there a need to specify a model. They are universal approximators meaning given enough neurons and processing time; we can model any relationship between the input and output variable to any degree of precision, making them ideal for predictive modeling. But this does not mean that they are universally best all the time.

The neural network is based on a brain's biological activity model; it can find a nonlinear relationship. It can discover interactions and it can do those things automatically. Neural networks perform very well with high-dimensional data. So, the significant advantage of a neural network is its flexibility.

A neural network is a black box describing the relationship between your target and the input variables. Neural networks work best with data with a strong signal when the signal-to-noise ratio is high. The network "learns" and updates its model iteratively as more data are fed into it. The most considerable disadvantage or objection is their lack of interpretability, called the black box objection. Another significant disadvantage, though, is overfitting which can be reduced by stopped training or weight decay.

7.1.1   How Do Neural Networks Learn?

Neural networks mostly perform supervised learning tasks, initiating knowledge from datasets where the proper answer is specified in advance. The networks then learn by tuning themselves to find the correct answer independently, increasing the accuracy of their predictions. Thus, the network compares initial outputs with a provided correct answer or target.

For a neural network to learn, there must be an element of feedback involved. It will be slightly different from what we saw for decision trees, where complexity changes by adding more branches and more leaves to become more significant. It is also different from regression models, where we can also add more predictors. With neural networks, we can make changes to the architecture that will change the number of weights. Learning is the process of optimizing complexity; this involves controlling the magnitude of the weights. We are not physically changing the model; instead, we are going through an iterative learning process.

This learning process is like finding the parameter estimates in a nonlinear regression model. Its primary goal is to estimate the weights and bias values by minimizing function loss. We will want to avoid what we call bad global minima and overfitting during this learning process. The key to avoiding overfitting and bad global mimima is the starting point for optimization. These will be the initial weight estimates for where the optimization begins. We specify some starting points that involve reading the input X and finding the answer Y by tuning the value W (Synaptic weight) for the optimization. We provide the network with training data. Then the neural network will control the weights.

The training uses iterations of data that are processed within the network. The more factors involved for the network to process, the more nodes (or units) are needed, and usually, the longer it will take to train the network. Then the neural network will control the weights that may be initialized randomly; this serves the process of symmetry-breaking and produces improved accuracy. Alternatively, we can also use He et al. (2015) a controlled initialization where the weights are modified, considering the size of the previous layer. This process aids in achieving a global minimum of the cost function faster and more efficiently. The weights are still random but vary in range depending on the size of the previous layer of neurons, multiplied by two. We do not want to let the weights get too big because our model will not generalize well; it will overfit. Weight decay and early stopping are the two primary methods to help avoid overfitting.

The weights get updated through propagation, and the algorithm learns typically by a backpropagation feedback process or gradient descent. This encompasses assessing the output the network generates with the created output and using the *difference* to modify the weights of the connections between the units in the network, working from the output units through the hidden units to the input units going backward.

In time, backpropagation initiates learning, thus reducing the difference between actual and intended output to the point where the two coincide. Hence, the network figures things out exactly as it should. Once the network has been trained with enough learning examples, it reaches a point where you can present it with an entirely new set of inputs it's never seen before and see how it responds.

7.2 The Architecture of Neural Networks

There were different attempts to structure the brain for the last sixty years. Ten years ago, we had some amount of semblance in terms of simulating the brain. The brain has 86 billion neurons (processing units). Then, if we connect them all in some meaningful form and replicate the brain's structure, we have neural networks that can be used in deep learning, working similarly to our brains' biological neural networks.

A typical artificial neural network (ANN) architecture known as multilayer perceptron (MLP) contains layers composed of neurons and their connections. MLP employs a supervised learning technique called backpropagation for training.

The main pieces of a neural network are the Input layer, Hidden layer (there can be several), and Output layer. In Fig. 7.1, we present an example of a neural network with an input layer, two hidden layers, and an output layer.

The basic unit of computation, the neuron, is often called a node. Each neuron in the network transforms data using a series of computations. The input layer is the layer that contains what we are feeding into our algorithm. The hidden layer is where most of the work happens. This layer is split into the weighted sum calculation and the activation calculation. This is where the neuron multiplies an initial value by some weight, sums results with other values coming into the same neuron, adjusts the resulting number by the neuron's bias, and then normalizes the output with an activation function. This process is repeated until the final output layer can provide scores or predictions related to the classification task at hand. Lastly, a cost function (error function) is used to calculate how well the algorithm did.

7.2.1 Terminology

A neural network always has a single input and output layer, but it can have multiple hidden layers. The optimal number of hidden layers is hard to know, may involve some trial and error, and maybe problem-specific or determined empirically (Principe et al., 2000).

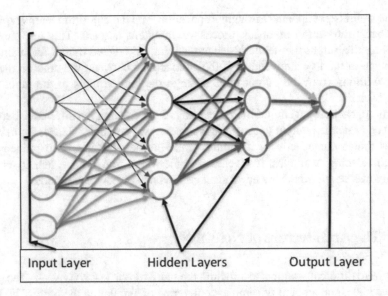

Fig. 7.1 Schematic neural network diagram

The basic terminology used is listed below.

– The input layer, the leftmost layer in Fig. 7.1, is the raw observations. They have no adjustable parameters (weights). They simply pass the positive or negative input to the next layer. We use X to refer to this input layer.
– Hidden layers are the middle layers where the neurons are neither inputs nor outputs; this is where a sigmoidal transformation of the weighted and summed input activations occurs.
– The output layer is the rightmost layer that shapes and combines the nonlinear hidden layer activation values.
– Nodes—circles are the nodes.
– Weights (like coefficients, subject to iterative adjustment)—w_{ij} displayed on arrows are weighted.
– Bias values (they are not subject to iterative adjustment)—B stands for bias. The bias is a neuron-specific number that changes the neuron's value once all the connections are processed.
– Weighted sum ($\sum w_{ij}$), also known as a *dot product,* is used to compute a value specified as the z value.
– Activation function: is the function that decides whether the neuron should fire and with what intensity. It ensures values passed on lie within a tunable, expected range. It uses a *sigmoid* function with a *parameter* of z.
– A *cost function (error function)* rates how good the neural network did as a whole.

Neural networks predict cases using an equation involving the input variables (Fig. 7.2). The prediction formula is used to predict new cases similar to regression but with a flexible addition. This addition enables a properly trained NN to model

Fig. 7.2 Neural network
prediction formula

Neural Network Prediction Formula
$\hat{Y} = f(\hat{w}_{11}T_1 + \hat{w}_{12}T_2 + \cdots + \hat{w}_{1n}T_n + b_1)$
$\hat{Y}$ is the prediction weight estimate
$\hat{w}_{11}$ is the weight estimate
T_1 is the hidden unit
b_1 is bias estimate
f is activation function

virtually any association between input and target variables. As with regression, the predictions can be decisions, rankings, or estimates.

7.3 The Mathematics Behind Neural Network

To help understand neural networks, we need some knowledge of college Math, basic calculus, and linear algebra. Functions can be different with different NN models.

7.3.1 The Common Activation Functions

An activation function is an essential feature of an artificial neural network; it decides whether the neuron should be activated or not. The activation functions convert the weighted sum of input signals of a neuron into the output signals. The activation calculation uses a sigmoid function with output signals serving as inputs to the next layer.

The standard activation functions used for neural networks are divided into two types:

1. Linear activation function where $f(x) = x$ and the range in between $-\infty$ and $+\infty$.
2. Nonlinear activation functions are the most used activation functions. A neural network without any activation function is simply a linear regression model.

The most commonly used nonlinear activation functions are stated in Table 7.1.

The Rectified Linear Unit (ReLU) function is another popular nonlinear activation function in the deep learning field. ReLU (Jarrett et al., 2009; Glorot et al., 2011) does not saturate like sigmoidal functions; thus, it is easier to quickly train deep neural networks by alleviating the difficulties with weight initialization and vanishing gradients. It can be used with Hidden layers.

The activation functions that we want to concern ourselves with are the activation functions in the hidden layer. For example, the hyperbolic tangent.

Table 7.1 Nonlinear activation functions

Activation function	Formula	Range	Usually used in
Sigmoid	$f(x) = \dfrac{1}{1+e^{-x}}$	Between 0 and 1	The output layer of binary classification
Hyperbolic tangent/Tanh	$f(x) = \dfrac{e^x - e^{-x}}{e^x + e^{-x}}$	Between −1 and 1	Hidden layers

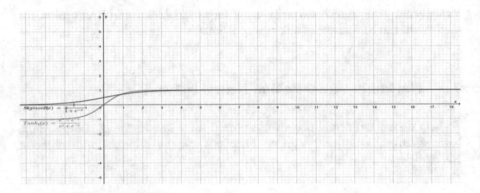

Fig. 7.3 Sigmoid and Tanh functions

7.3.2 *Limit of Functions*

In mathematics, a limit is a value that a function (or sequence) "approaches" as the input (or index) "approaches" some value. For instance, if z approaches a negative number, then the limit of e^{-z} is $+\infty$. Now, if z approaches $+\infty$, then the limit of the sigmoid function, $f(z) = \dfrac{1}{1+e^{-z}}$ is 1 and the limit of Tanh function, $f(z) = \dfrac{e^z - e^{-z}}{e^z + e^{-z}}$ is 1 where if z approaches $-\infty$, then the limit of the sigmoid function $f(z) = \dfrac{1}{1+e^{-z}}$ is 0 and the limit of Tanh function, $f(z) = \dfrac{e^z - e^{-z}}{e^z + e^{-z}}$ is −1, as shown graphically in Fig. 7.3:

7.3.3 *Chain Rule*

The algorithm is used to effectively train a neural network through a method called chain rule. What is a chain rule?

In mathematics, the composition of functions is when one function is inside of another function. Let f, g be functions. Then, we can produce the function h such that $h(x) = g(f(x))$ which is denoted by $g \circ f$, defined by $(g \circ f)(x) = g(f(x))$. For instance, consider $f(x) = 3x - 5$ where the domain and the range is the set of real numbers R and $g(x) = x^2$ where the domain is R, and the range is the set of the positive real number R^+. Then, the composition function, $h(x) = g(f(x)) = g(3x - 5) = (3 x - 5)^2$, has domain R and range R^+. The Chain rule provides a technique to calculate the derivative of a composite function:

$\dfrac{dh(x)}{dx} = \dfrac{dg(f(x))}{df(x)} \times \dfrac{df(x)}{dx}$. Let $f(x) = u = 3x - 5$. Then, Fig. 7.4 displays the-complete formula.

For instance, the error function (cost function) depends on the activation function, which depends on z, and z depends on w. For example, the derivative of the error function concerning w_{11} is:

$$\frac{dE}{dw_{11}} = \frac{dE}{d\sigma} \times \frac{d\sigma}{dz_1} \times \frac{dz_1}{dw_{11}} \qquad \text{and} \qquad \text{the} \qquad \text{error} \qquad \text{function} \qquad \text{formula} \qquad \text{is}$$

$$E_{total} = \frac{1}{2}(\text{target} - \text{output})^2$$

7.3.4 Working of Neural Network

Information is fed to the input layer; the interconnections between the two layers assign weights to each input randomly. Weights can be positive (if one unit excites another) or negative (if one unit suppresses or inhibits another). The higher the weight, the more influence one unit has on another. The neurons perform a linear transformation using the weights and biases: $x = (\text{weight} * \text{input}) + \text{bias}$. An activation function is applied to the result.

$$Y = \text{Activation}\left(\sum(\text{weight} * \text{input}) + \text{bias}\right)$$

Finally, the activation function's output moves to the next hidden layer, and the same process is repeated. This forward movement of information is known as forward propagation. We follow with an example that contains two hidden layers to explain this principle in the subsection below.

$$\frac{dh(x)}{dx} = \frac{dg(f(x))}{df(x)} \times \frac{df(x)}{dx} = \frac{dg(u)}{du} \times \frac{du}{dx} = 2u^{2-1} \times 3 x^{1-1} = 2(3x-5) \times 3 = 6(3x-5) = 18x-30$$

$$\text{We use } \frac{dg(u)}{du} = nau^{n-1} \text{ and } \frac{df(x)}{dx} = nax^{n-1}$$

Fig. 7.4 Chain rule formula

7.3.4.1   Forward Propagation

With forward propagation, we check what the NN predicts for the first training with the initial weights and bias and then compute the z, the weighted sum of activation and bias.

Let us illustrate how this works and assume that we have data with binary classification. We have input x_1 with weights (w_{11}, w_{21}, w_{31}), x_2 has (w_{12}, w_{22}, w_{32}), etc. Here, our goal is to show how to represent the weights at different layers and compute them. Remember that the neuron performs both summation and activation. Figure 7.5 depicts the connection between layers.

Inside each neuron, there are two actions. First, in computing the linear summation, for instance, in the first neuron in the hidden layer one, we have this summation denoted by $z_1 = w_{11} \cdot x_1 + w_{12} \cdot x_2 + w_{13} \cdot x_3 + w_{14} \cdot x_4 + b_1$, and then, there is some function, for instance, σ (activation function), which is not linear, which implies $\sigma(z_1) = s_1$.

The same process happens in each neuron, but we vary the activation functions in hidden layer neurons, not in the output layer.

s_1 is the output in neuron one from hidden layer one. Similarly, we can have s_2 and s_3, respectively, from the hidden layer one. These outputs become inputs to the hidden layer 2. Then, we have $\sigma'(T_1)$ the output from neuron one in the hidden layer 2 where $T_1 = w'_{11}s_1 + w'_{12}s_2 + w'_{13}s_3 + b_2$ is the summation in the neuron one in the hidden layer 2, the weights are $(w'_{11}, w'_{12}, w'_{13})$ and σ' is the activation function.

Hence, $\sigma'(T_1)$ is the output from neuron two in hidden layer 2, and similarly, we have $\sigma'(T_2)$ is the output from neuron two in the hidden layer 2 supplies inputs to the output layer. Finally, the output layer is represented in Fig. 7.6 below.

Therefore, the predicted value is

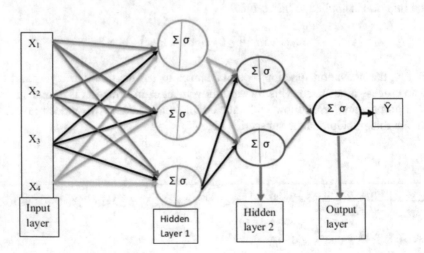

Fig. 7.5 Connections between layers

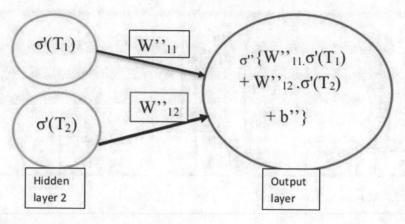

Fig. 7.6 Output layer

$$\hat{Y} = \sigma''\left\{ W_{11}'' \cdot \sigma'(T_1) + W_{12}'' \cdot \sigma'(T_2) + b'' \right\}$$

We presented above how the neuron works for a simple network, four layers, and six neurons. The computations are harder with an extensive network of 15 layers with 150 s of neurons, for instance. It is almost impossible to do an efficient manual calculation. If our dataset has m entries with nx features each, we can create matrices to represent neural networks, generating concise and precise computations.

Let us illustrate briefly how this works. We create a matrix for each layer X (represents the inputs), W (represents the weights), and Z represents the outputs from this layer where b is a vector, then we have $Z = W * X + b$.

We present the matrix version calculations, which are performed for a whole layer of the neural network. We consider what is happening inside a single unit and vectorize it across the hidden layer to combine those calculations into matrix equations. Then, we have the following matrices, as shown in Fig. 7.7.

Hence, we can see that many programming languages support matrices because matrices in the neural network perform a smooth and efficient calculation.

7.3.5   Backward Propagation

Backpropagation (aka gradient descent) helps to assess a given input variable's impact on a network output. The knowledge gained from this analysis is represented in rules.

Backpropagation is spreading the total loss back into the neural network to discern how much of the loss every node is responsible for. Consequently, the weights update to minimize the loss by giving the loss nodes higher error rates, lower weights, and vice versa. It consists of two steps. First, it calculates gradients of the loss/error function, and then it updates existing parameters in response to the

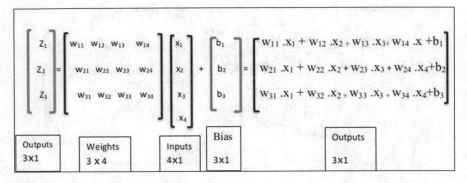

Fig. 7.7 Matrices with calculations

gradients, which is how the descent is done. This cycle is repeated until reaching the minima of the loss function.

We usually optimize complexity for neural networks through the learning process and the numerical methods used to search for updates in weight estimates. We first try to find a good set of parameters that minimizes the error and performs well, avoiding overfitting. Then we apply the chain rule to differentiate composite functionssto a single neuron. Backpropagation uses gradient descents, $\dfrac{dE}{w_{11}}$ and $\dfrac{dE}{w_{12}}$ the

vectors of partial derivatives of the error function concerning the weights determine an appropriate set of weights. As more layers using certain activation functions are added to neural networks, the gradients of the loss function approach zero, making the network hard to train. The idea is that we want to get to the *minima* (lowest point) of the surface to minimize cost through gradient descent.

Gradient descent is an iterative optimization algorithm used to minimize a loss function that describes how well the model will perform given the current set of parameters (weights and biases). Using Calculus, Gradient descent is used to find the best set of parameters corresponding to the minimum value of the given cost function. Mathematically, the "derivative" technique is fundamental to minimize the cost function because it helps to get the minimum point. The derivative refers to the function's slope at a given point. We need to know the slope to know the direction (sign) to move the coefficient values to get a lower cost on the next iteration.

Training a neural network with gradient descent requires the calculation of the gradient of the error function E depending on the output, weight, and the activationfunction, where the error function formula is: $E_{total} = \dfrac{1}{2}(\text{target} - \text{output})^2$.

To illustrate, consider the activation function, σ, sigmoid, and the following summation in neuron one: $z_1 = w_{11} \cdot x_1 + w_{12} \cdot x_2 + b_1$.

Then we calculate the partial derivative for $E(\sigma)$.

$$\frac{dE}{dw_{11}} = \frac{dE}{d\sigma} \times \frac{d\sigma}{dz_1} \times \frac{dz_1}{dw_{11}} \text{ and } \frac{dE}{dw_{12}} = \frac{dE}{d\sigma} \times \frac{d\sigma}{dz_1} \times \frac{dz_1}{dw_{12}}.$$

Lastly, to minimize the error for each output neuron and the network, we calculate the derivative of the error concerning weights. At the base of any minima, the gradient descent is zero. A local minimum is called so when the loss function's value is minimum at the point in a local region. Whereas a global minimum is the value of the loss function is minimum across the entire domain. Now, having changed the weights concerning the error, we update the weights using gradientdescent $\dfrac{dE}{dw_{11}} / \dfrac{dE}{dw_{12}}$. Hence, we have the new weights as follows:

$$w_{11}\text{new} = w_{11} \ \text{old} - \eta \frac{dE}{dw_{11}} \ \text{and} \ w_{12}\text{new} = w_{12} \ \text{old} - \eta \frac{dE}{dw_{12}}, \text{Where } \eta \text{ is a}$$

small,positive parameter (known as the learning rate), this process continues for all samples until the error reaches a minimum value. The small changes Δw_j (delta) in the weights and Δ_b in the bias will produce a small change Δoutput in the neuron's output, then, σ is a smooth function.

$$\Delta_{output} \approx \sum_j \frac{\sigma_{output}}{\sigma_{w_j}} \Delta_{w_j} + \frac{\sigma_{output}}{\sigma_b} \Delta_b$$

where $\dfrac{\sigma_{output}}{\sigma_{w_j}}$ and $\dfrac{\sigma_{output}}{\sigma_b}$ denote partial derivatives of the output concerning Δw_j and Δ_b, respectively. Hence, Δ the output is a linear function of the changes Δw_j and Δ_b in the weights and bias. Remark, when we use a different activation function, this changes the particular values for the partial derivatives.

7.4   Vanishing and Exploding Gradient

A difficulty that we face when training deep Neural Networks is vanishing or exploding gradients.

When training a deep neural network with gradient-based learning and back-propagation, if the derivatives are significant, the gradient will increase exponentially as we propagate down the model until it eventually explodes. This is called the problem of exploding gradient. The explosion occurs exponentially by repeatedly multiplying gradients through the network layers with values larger than 1.0.

Vanishing gradients is a particular problem with recurrent neural networks. If the derivatives are small, the gradient will decrease exponentially as we propagate through the model until it eventually vanishes, which is the vanishing gradient problem. To deal with vanishing, we can use other activation functions, such as ReLU, which doesn't cause a small derivative. Residual networks are another solution, as they provide residual connections straight to earlier layers.

It is important to note that both problems accumulate large and small derivatives, resulting in the model being incapable of active learning, creating an unstable network.

7.5   Demonstration for Neural Networks

7.5.1   Concrete Compressive Strength Example 1: SAS Visual Analytics

Inside a Concrete Solution company, a neural network could be used for quality control. Let's say you're producing concrete; you could measure the final compressive strength in various ways (its flowability, strength, or whatever), feed those measurements into your neural network as inputs, and then have the network decide whether to accept or reject the mix.

We are using the same example from the previous chapters. The concrete compressive strength is a highly nonlinear function of age and ingredients, and we will use it as our predictor variable (refer to Appendix D). We are still looking at answering what determines the compressive strength of concrete, our target response.

As presented in Fig. 7.8, we created a pipeline with default settings and autotuned settings for this demonstration. We will explain the autotuned results as they provided the most significant results.

Under the data roles, we select the target measure or response variable, in this case, the concrete compressive strength and interval target, and add the predictors. With this type of target variable, we also need to select the distribution and activation function. If you select the gamma or Poisson distribution, an exponential activation function is used. If you select a normal distribution, the identity activation function is used by default. We did not have missing variables.

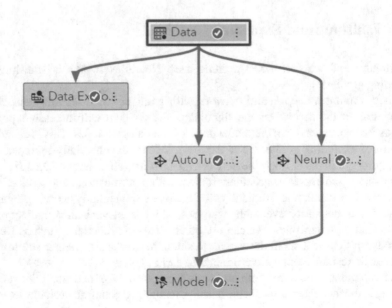

Fig. 7.8 NN pipeline

We used a 60–40 split. From the options tab, we used Autotune to find optimal values for model hyperparameters. The autotuning select the Number of hidden layers, Number of neurons, L1, L2, and Maximum iterations values that produce the best model. Also, since data partitioning is applied, then the Auto-stop method is also autotuned.

- L1 specifies the L1 regularization parameter. It penalizes the absolute value of weight. The weights shrink by a constant amount toward 0.
- L2 specifies the L2 regularization parameter. The weights shrink by an amount that is proportional to weights.

Figure 7.9 presents the NN output for our model.

The summary bar appears at the top of the canvas with the model type, Name of the response variable, in this case, Concrete Compressive Strength, Model evaluation criteria, the validation ASE of 27.1563, meaning that, on average, our model has produced predictions that are off from the actual value by 27.15. One of the goals of our model tuning steps would be to reduce this value to predict the compressive strength with a higher degree of accuracy and produce a better fit model. The number of observations used to build the model, in this case, is 1030 observations.

Figure 7.10 highlights a detailed description of our NN model, starting with the number of observations used to the total number of neurons (20). We can see that we have eight input neurons and one hidden layer.

From an architecture perspective (as presented in Sect. 7.2), the first plot on the left side (Fig. 7.9) is the network diagram that displays the input nodes, hidden nodes, connections, and output nodes of NN. The circle's size represents the absolute value of estimated weights; color indicates whether the weights are positive or negative. The width of the line between two nodes indicates the strength. For

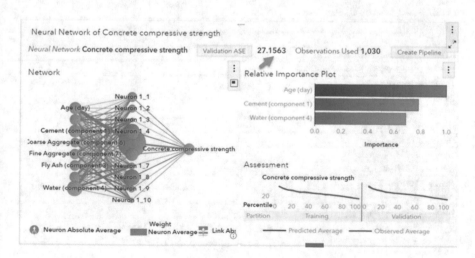

Fig. 7.9 NN output

example, looking at the age node, 1–5 neuron (Age) has the highest weight among others depicted by the blue color circle.

There are several charts and plots to help evaluate the model's performance. One of these plots is the *Network Diagram: Top 200 Weights*, which presents the final neural network structure for this model, including the hidden layer and the hidden units (Fig. 7.11).

To highlight what happens within the hidden layers and connections (Fig. 7.9), we show the sigmoid line between the input—fly_ash and H_1_10; we can see the weight of this sigmoid 0.0155 as well (Fig. 7.11).

Looking at the relative importance plot in the upper right side of Fig. 7.9, we can notice age, cement, and water are the essential components for the mix.

The assessment plot, cumulative lift (bottom right Fig. 7.9), helps to determine how well the model fits the data and how better the model is than the random events. The gap between training and validation accuracy is negligible, indicating a good model fit. The larger the gap, the higher the overfitting.

For instance, we can create another plot, the Iteration plot (Fig. 7.12), that displays the value of the objective/loss function at each iteration in the network building process. After many iterations of about 50, the objective function flattens.

7.5.2   Demonstrations of Neural Network: Aviation Example

In airplanes, we may use a neural network as a basic autopilot, with input units reading signals from the various cockpit instruments and output units modifying the plane's controls appropriately to keep it safely on the course (Woodrow 2020).

Neural network - Concrete compressive strength 1 Supplement 1

Description	Value
Model	Neural Net
Number of Observations Used for Training	618
Number of Observations Read for Training	618
Target/Response Variable	Concrete compressive strength
Number of Neurons	20
Number of Input Neurons	8
Number of Output Neurons	1
Number of Hidden Neurons	11
Number of Hidden Layers	1
Number of Weight Parameters	99
Number of Bias Parameters	12
Architecture	MLP
Number of Neural Nets	1
Objective Value	144.4977
Mean Squared Error for Validation	255.0312

Fig. 7.10   Description summary of NN model